# Strategic Planning Decisions in the High Tech Industry

Dilek Cetindamar · Tugrul Daim
Berna Beyhan · Nuri Basoglu
Editors

# Strategic Planning Decisions in the High Tech Industry

Springer

*Editors*
Dilek Cetindamar
School of Management
Sabanci University
Istanbul
Turkey

Berna Beyhan
Science, Technology and Policy Center
METU
Ankara
Turkey

Tugrul Daim
Engineering and Technology Management
Portland State University
Portland
USA

Nuri Basoglu
Management Information Systems
Boğaziçi University
Istanbul
Turkey

ISBN 978-1-4471-4886-9 ISBN 978-1-4471-4887-6 (eBook)
DOI 10.1007/978-1-4471-4887-6
Springer London Heidelberg New York Dordrecht

Library of Congress Control Number: 2013931272



Printed on acid-free paper

Springer is part of Springer Science+Business Media (www.springer.com)

# Contents

# Chapter 1
# Strategic Planning Decisions: An Overview

**Tugrul Daim, Dilek Cetindamar, Berna Beyhan and Nuri Basoglu**

Who would have thought 10 years ago that Apple would be a phone company or even a music company? As we are witnessing another technological discontinuity, once again we realize the importance of making informed decisions in the high-tech industry. This book provides a snapshot of the tools being developed to address the uncertainty induced by technological discontinuities.

This chapter precisely explores specific decisions with different applications documented in the literature. We will review three major topics that transform strategic planning into actionable decisions in technology-driven organizations: technology planning, technology evaluation, and technology diffusion (Fig. 1.1).

## 1.1 Technology Planning

Everything starts with a plan. Phaal et al. [1] provide a very simple but effective tool for technology planning: technology roadmaps. The use of roadmaps, especially technology roadmaps, is widely used in the industry and in government

T. Daim (✉)
Engineering and Technology Management, Portland State University, SW Fourth, Portland 97201, USA
e-mail: tugrul@etm.pdx.edu

D. Cetindamar
Faculty of Management, Sabanci University, Tuzla, 34956 Istanbul, Turkey

B. Beyhan
Institute for Innovation and Entrepreneurship, School of Business, Economics and Law, Gothenburg University, Gothenburg, Sweden

N. Basoglu
Management Information Systems, Bogazici University, Hisar Campus, Bebek, 34342 Istanbul, Turkey
e-mail: basoglu@boun.edu.tr

D. Cetindamar et al. (eds.), *Strategic Planning Decisions in the High Tech Industry*, DOI: 10.1007/978-1-4471-4887-6_1,

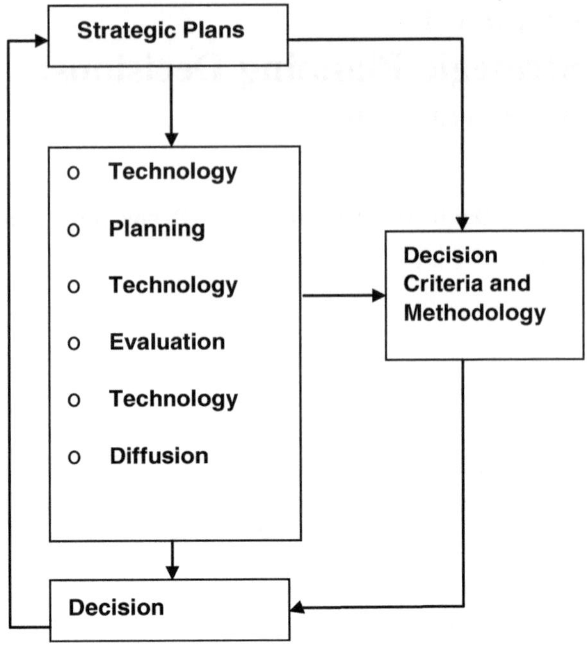


**Fig. 1.1** Strategic planning decision framework in the high-tech industry

organizations. They are used for the basic technology planning function: linking organizational strategic goals to research and development investment decisions while also communicating these linkages visually [2]. The process starts with identification of the internal and external forces impacting the goals and strategies of the organization as well as documentation of its vision. Next, organization's current capabilities in terms of products or services are documented. The process helps identify the gaps between the current capabilities and the vision of the organization at product/service, technology, and research and development levels by linking all of these levels. Rinne [3] expanded the concept to multiple dimensions to help with identifying those opportunities which may not be identified as goals yet.

Mapping internal and external activities provides an insight into forces at play for any given sector. Phaal et al. [4] introduced a framework for mapping. Technology planning uses several inputs including the assessment of technologies, which we will discuss in the following section. Thorn et al. [5] demonstrated a practical application of technology valuation. Martin and Daim [6] building upon Gerdsri's [7] approach of technology development envelope, introduced valuation of technologies for service-oriented sectors. Gerdsri et al. [8, 9] emphasized the importance of getting organized for the activity of technology planning including functions ranging from technology roadmapping to prioritization. They imply that you should match the tools with an organizational structure including project

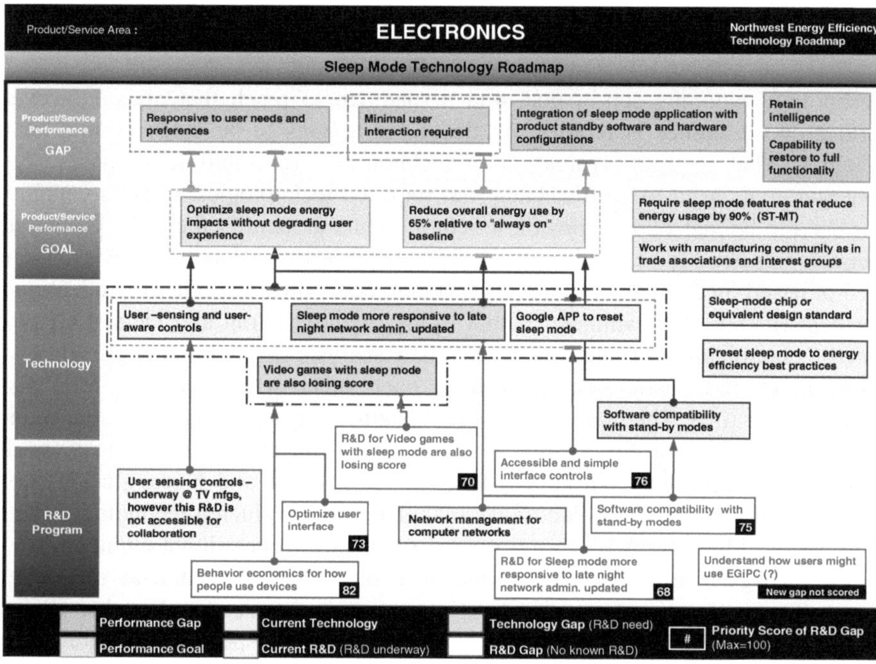


**Fig. 1.2** An extract from the US NW energy efficiency technology map

teams. Fenwick et al. [10] argued that technology roadmapping is the platform that integrates all planning activities ranging from marketing to engineering. They matched different planning tools to different components of the technology roadmapping process. Abe [11] also demonstrated a similar integrated approach. Lee and Park [12] presented how we can customize our approach. Petrick and Echols [13] also showed us how we can make better product development decisions with approaches like roadmapping.

Technology planning is not required solely in the product or service organizations but in many diverse sectors such as energy and government. According to Amer and Daim [14] roadmaps are widely used in renewable energy sector. Several studies have provided applications which can easily be duplicated in practice. These include applications in bio fuels [15], electricity (Fig. 1.2, [16, 17]), government policies [18], automotive [19], semiconductors [20], materials [21], national planning and foresight [22–24], general science and technology [25], and corporate foresight [26].

Many emphasize the importance of identification of discontinuities which really makes it challenging to manage technologies. References [27–29] provide varying methods to identify discontinuities.

## 1.2 Technology Evaluation

Technology planning sets a goal and a roadmap to reach that goal. However, further due diligence is required to detail the actions to be made around technologies. That requires evaluation of technologies for multiple purposes. The strategic goals need to be translated into required technologies. Current capabilities need to be matched against these requirements to identify the gaps. Candidate technologies need to be identified and evaluated to address these gaps. Evaluations are not based on technical criteria only; social, economic, and other organizational perspectives are also evaluated with varying methods. Tran and Daim [30] provided a classification of these methods. Their analysis showed that applications in government and business differed from each other while methods used also changed over time. The main reason is that business organizations are interested in maximizing their profits while government organizations have many goals including society and environment focused ones. Daim and Kocaoglu [31] discovered that businesses in electronics manufacturing evaluated technologies with the goal of either increase in sales or improvement of operational efficiency.

Technology evaluation is not based on technical aspects only, as mentioned above. There are several non-technical perspectives that need to be taken during evaluation. Many of them, including social or political perspectives, require expert judgment and quantification of those judgments. Methods addressing these perspectives include analytical hierarchy process, analytical network process, and hierarchical decision models (Fig. 1.3). Iskin et al. [32] applied analytic networking process to evaluate the impact of economic and other non technical aspects on renewable energy. Daim et al. [33] integrated fuzzy delphi and analytic hierarchy processes together to evaluate technologies across multiple perspectives. Such models have been applied in evaluation of offshore wind turbine designs

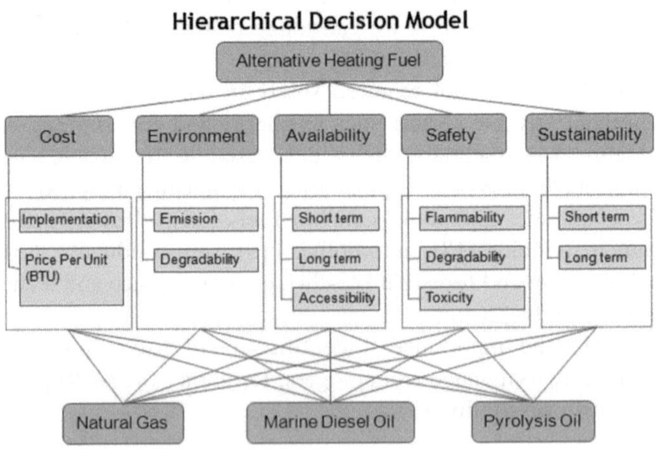


**Fig. 1.3** A hierarchical decision model for evaluating heating fuel

[34], renewable energy technologies in Pakistan [35]; energy resources in China [36], and renewable energy in the Pacific Northwest [37, 38].

Evaluation needs to be balanced with building portfolios since evaluation rates alternatives across different criteria while portfolios balance these criteria. Examples of such applications include hydroelectric power and storage portfolio [39] and renewable energy portfolio [40].

While it is important to understand the current capabilities and impact of technologies, foreseeing their future is also essential since naturally no one would like to invest in a technology which will be obsolete soon. Technology forecasting is the general name given to methods which help explore the evolution of technologies into the future. Methodologies range considerably in this area as well. Kim et al. [41] used regression analysis to forecast wireless technologies while Anderson et al. [42] used mathematical programming, specifically data envelopment analysis for the same purpose. As Daim et al. [43] emphasized, it is useful to use multiple methods so that further insight could be attained as we are really looking into the future which is not an easy task at all.

Exploring future requires intelligence on trends. Many indicators are used for this purpose; inventors [44], patents [45–48] and publications [49].

Growth curves, or "S" curves as they are commonly called, are also widely used for forecasting technologies [50, 51].

## 1.3 Technology Diffusion

One of the critical elements in planning and evaluating technologies is the comprehension of potential transfer and diffusion of technologies. Many studies explored technology transfer in different organizations. Anderson et al. [52] and Kim et al. [53] explored university technology transfer and identified efficiency differences among universities and potential causes for that. Daim and Kocaoglu [54] studied the technology acquisition in different geographies and identified commonalities and differences in approaches.

While acquisition and transfer are planned technology transitions, there are other dynamics that we cannot control 100 % but try to influence. The main dynamic is the adoption of technologies by users. Many studied this in different sectors and provided very practical tools. Sectors and application include wireless devices in health care [55], diet software [56] online flight reservation systems [57], and enterprise resource planning (Fig. 1.4, [58, 59]).

## 1.4 Book Content

The chapters of this book provide examples of technology planning, evaluation, and diffusion.

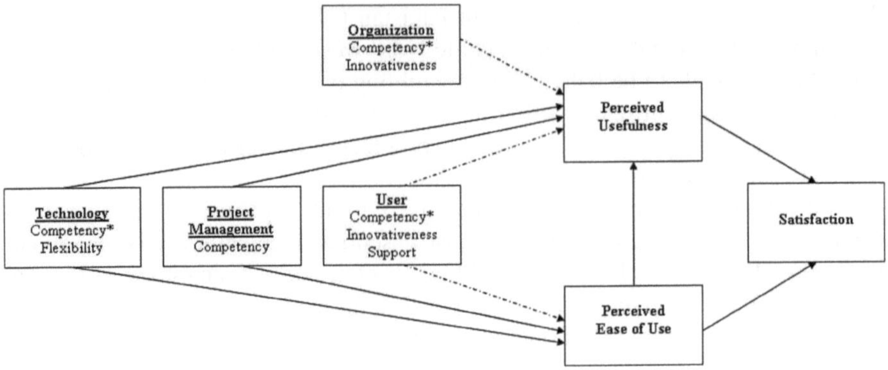


**Fig. 1.4** Technology adoption model for enterprise resource planning [58, 59]

## *1.4.1 Technology Planning*

Planning approaches vary a lot, and different researchers [25] have studied many of them.

Beyhan and Cetindamar explore the fundamentals of the technology management (TM) literature generated in developing countries using citation and co-citation analyses and answer the question of whether the intellectual bases of the TM literature created by authors in developing countries diverge from those of the global TM literature.

Amezcua-Martinez and Guemes-Castorena provide a strategic foresight methodology to identify technology trends and business opportunities. The authors suggest that this model is expected to respond to the following strategic basic foresight questions: what it is that is being sought, what the technologies are, and the direction that needs to be taken.

Ardilio et al. take us into how strategic planning is done for research institutions. They introduce a method for the identification of actual and future market needs and their transformation to the strategic technology planning by using graph theoretical approaches.

Kokshagina et al. explore how we can guide the development in the situation of double "unknown" of technology and market to enable discontinuous radical innovation. The model allows creating a space of working on highly uncertain object without fixing concrete market functionality at the beginning and allowing flexible change on product configuration.

Kajikawa et al. [25] integrate risk analysis and scenario planning for technology roadmapping for energy technologies. This approach includes identification of technological, economic, environmental, and social risks and constraints and applies the scenario planning concept. Inputs from more than 40 experts in energy sector were collected and assessed to synthesize an integrated perspective on energy technologies, consistently describe inherent risks, and quantify expected outcomes.

### 1.4.2 Technology Evaluation

Cases for several evaluation concepts are provided by Dereli and Altun, Cunningham and van der Lei, Farrukh et al., Bunzel et al., and Krcelic et al.

Dereli and Altun demonstrate evaluation and prioritization of candidate technologies with respect to their innovation potentials. Their chapter presents a novel process, namely "quick innovation intelligence process", which aims to evaluate and prioritize candidate technologies according to their innovation potentials by considering commercialization, imitation, and trendiness factors together.

Cunningham and van der Lei introduce a decision-making approach for a new technology leveraging exchange modeling, which incorporates actor preferences and assumptions about the play of the game, to better examine the resulting preferences of groups.

Farrukh et al. attempt to evaluate and communicate the value of technology. They propose a "technology marketing" process to seek out and to identify suitable application opportunities for a promising technology, and to engage with potential partners using a consultative sales framework.

Bunzel et al. introduce an ontology-based technology model for the use in the early stage of product development. The authors demonstrate the benefits by applying it to nanotechnology.

Krcelic et al. attempts to reveal operational improvement of fixed network planning process derived from Six Sigma knowledge, methods, and tools. They present a model for quantification of benefits for given investment in access network and possible ways to identify the best locations for investment based on data-driven business conclusions drawn.

### 1.4.3 Technology Diffusion

Finally, chapters by Ortiz-Gallardo et al. and Magnusson and Rickne provide examples of different approaches exploring technology diffusion.

Ortiz-Gallardo et al. present a conceptual framework that provides an overview of the key aspects that have influence on the performance of technology acquisitions by collaboration. Their findings suggest that five types of factors are relevant; contextual issues, technology uncertainty, strategic alignment, structural match, and coordination capability.

Magnusson and Rickne link the approach of multilevel governance with that of technological innovation systems. They perceive the concept of technological innovations system as a missing link between niche and regime. By applying their framework to the case of electric vehicles, they show that it is crucial to understand the interplay of niche technologies.

## References


1. Phaal R, Farrukh CJP, Probert DR (2004) Technology roadmapping—a planning framework for evolution and revolution. Technol Forecast Soc Chang 71(1–2):5–26
2. Phaal R, Muller G (2009) An architectural framework for roadmapping: towards visual strategy. Technol Forecast Soc Chang 76(1):39–49
3. Rinne M (2004) Technology roadmaps: infrastructure for innovation. Technol Forecast Soc Change 71(1–2):67–80
4. Phaal R, O'Sullivan E, Routley M, Ford S, Probert D (2011) A framework for mapping industrial emergence. Technol Forecast Soc Chang 78(2):217–230
5. Thorn V, Hunt F, Mitchell R, Probert D, Phaal B (2011) Internal technology valuation: real world issues. Int J Technol Manage 53(2–4):149–160
6. Martin H, Daim T (2012) Technology roadmap development process (TRDP) for the service sector: a conceptual framework. Technol Soc 34(1):94–105
7. Gerdsri N (2007) An analytical approach to building a technology development envelope (TDE) for roadmapping of emerging technologies. Int J Innov Technol Manage 4(2):121–135
8. Gerdsri N, Vatananan RS, Dansamasatid S (2009) Dealing with the dynamics of technology roadmapping implementation: a case study. Technol Forecast Soc Chang 76(1):50–60
9. Gerdsri N, Assakul P, Vatananan RS (2010) An activity guideline for technology roadmapping implementation. Technol Anal Strateg Manage 22(2):229–242
10. Fenwick D, Daim T, Gerdsr N (2009) Value driven technology road mapping (VTRM) process integrating decision making and marketing tools: case of internet security technologies. Technol Forecast Soc Chang 76(8):1055–1077
11. Abe H, Ashiki T, Akihiko S, Jinno F, Hiraku S (2009) Integrating business modelling and roadmapping methods—the innovation support technology (IST) approach. Technol Forecast Soc Chang 76(1):80–90
12. Lee S, Park Y (2005) Customization of technology roadmaps according to roadmapping purposes: overall process and detailed modules. Technol Forecast Soc Chang 72(5):567–583
13. Petrick IJ, Echols AE (2004) Technology roadmapping in review: a tool for making sustainable new product development decisions. Technol Forecast Soc Chang 71:81–100
14. Amer M, Daim T (2010) Application of technology roadmaps for renewable energy sector. Technol Forecast Soc Chang 77(8):1355–1370
15. Lamb A, Daim T, Leavengood S (2012) Wood pellet technology roadmap. IEEE Trans Sustain Ener 3(2):218–230
16. Daim T, Amer M, Brenden R (2012) Technology roadmapping: wind energy for Pacific NW. J Cleaner Prod 20(1):27–37
17. Daim T, Oliver T (2008) Implementing technology roadmap process in the energy services sector: a case study of a government agency. Technol Forecast Soc Change 75(5):687–720
18. Yasunaga Y, Watanabe M, Korenaga M (2009) Application of technology roadmaps to governmental innovation policy for promoting technology convergence. Technol Forecast Soc Chang 76(1):61–79
19. Daim T, Gerdsri N, Kockan I, Kocaoglu DF (2011) Forecasting powertrain technologies. J Transp Sys Eng Inf Technol 11(2):58–69
20. Walsh S, Boylan RL, McDermott C, Paulson A (2005) The semiconductor silicon industry roadmap: epochs driven by the dynamics between disruptive technologies and core competencies. Technol Forecast Soc Chang 72(2):213–236
21. Lee S, Kang S, Park YS, Park Y (2007) Technology roadmapping for R&D planning: the case of the Korean parts and materials industry. Technovation 27(8):433–445
22. Saritas O, Oner MA (2004) Systemic analysis of UK foresight results: joint application of integrated management model and roadmapping. Technol Forecast Soc Chang 71(1–2):27–65
23. Chan L, Daim T (2012) Are foresight studies generating the right innovation policies: case of BRIC. Futures 44(6):618–630

24. Daim T, Basoglu N, Saritas O, Dursun O, Gerdsri P (2009) A comprehensive review of Turkish technology foresight project. Foresight 11(1):21–42
25. Kajikawa Y, Usui O, Hakata K, Yasunaga Y, Matsushima K (2008) Structure of knowledge in the science and technology roadmaps. Technol Forecast Soc Chang 75(1):1–11
26. Rohrbeck R, Gemünden HG (2011) Corporate foresight: its three roles in enhancing the innovation capacity of a firm. Technol Forecast Soc Chang 78(2):231–243
27. Vojak B, Chambers FA (2004) Roadmapping disruptive technical threats and opportunities in complex, technology-based subsystems: the SAILS methodology. Technol Forecast Soc Chang 71(1–2):121–139
28. Galvin R (2004) A practitioner's update roadmapping: from sustainable to disruptive technologies. Technol Forecast Soc Chang 71(1–2):101–103
29. Kostoff R, Boylan R, Simons GR (2004) Disruptive technology roadmaps. Technol Forecast Soc Chang 71(1–2):141–159
30. Tran T, Daim T (2008) A taxonomic review of methods and tools applied in technology assessment. Technol Forecast Soc Change 75(9):1396–1405
31. Daim T, Kocaoglu D (2009) Exploring the roles of technology assessment in the competitiveness of US electronics manufacturing companies. Int J Technol Manage 48(1):77–94
32. Iskin I, Kayakutlu G, Daim T (2012) Renewable energy pricing and revenue management: a comparative analysis. Energy Econ 34(4):882–891
33. Daim T, Li X, Kim J, Simms S (2012) Evaluation of energy storage technology: Integration of Fuzzy Delphi Method, AHP and Fuzzy Consistent Matrix. Environ Innov Soc Trans 3:29–49
34. Daim T, Bayraktaroglu E, Estep J, Lim DJ, Upadhyay J, Yang J (2012) Optimizing the NW off-shore wind turbine design. Mathematical Comput Modeling 53(3–4):396–404
35. Amer M, Daim T (2011) Selection of renewable energy technologies for a developing county: a case of Pakistan. Energy Sustain Dev 15(4):420–435
36. Wang B, Kocaoglu D, Daim T, Yang J (2010) A decision model for energy resource selection in China. Energy Policy 38(11):7130–7141
37. Daim T, Cowan K (2010) Assessing renewable energy portfolio futures with multiple perspectives. Technol Soc 32(4):255–263
38. Daim T, Yates D, Peng Y, Jimenez B (2009) Technology assessment for clean energy technologies. Technol Soc 31(3):232–243
39. Cowan K, Daim T, Anderson T (2010) Exploring impact of technology development and adoption for sustainable hydroelectric power and storage technologies in the Pacific Northwest United States. Energy 35:4771–4779
40. Daim T, Kayakutlu G, Cowan K (2010) Developing Oregon's renewable energy portfolio using fuzzy goal programming model. Comput Ind Eng 59(4):786–793
41. Kim J, Daim T, Anderson T (2010) A look at the future of wireless technologies. Technol Anal Strateg Manage 22(8):925–943
42. Anderson T, Daim T, Kim J (2008) Forecasting wireless technologies with DEA. Technovation 28(9):602–614
43. Daim T, Grueda G, Martin H, Gerdsri P (2006) Forecasting emerging technologies: use of bibliometrics and patent analysis. Technol Forecast Soc Change 73(8):981–1012
44. He J, Fallah MH (2009) Is inventor network structure a predictor of cluster evolution? Technol Forecast Soc Chang 76(1):91–106
45. Järvenpää HM, Mäkinen SJ, Seppänen M (2011) Patent and publishing activity sequence over a technology's life cycle. Technol Forecast Soc Chang 78(2):283–293
46. Daim T, Iskin I, Li X, Zielsdorff C, Bayraktaroglu E, Dereli T, Durmusoglu A (2012) Development of a patent alert system for wind rotor technologies. World Patent Inf 34(1):37–47
47. Harell G, Daim T (2009) Forecasting energy storage technologies. Foresight 11(6):74–85
48. Daim T, Kennedy L, Choothian W, Ploykitikoon P (2008) Forecasting the future of data storage: case of hard disk drive and flash memory. Foresight 10(5):34–49

49. Behkami N, Daim T (2012) Methodological analysis of health information technology (HIT) diffusion research to identify gaps and emerging topics in literature. Technol Forecast Soc Chang 79(3):498–508
50. Daim T, Harell G, Hogabaum L (2012) Forecasting renewable energy production in the United States. Foresight 14(3):225–241
51. Lamb A, Anderson T, Daim T (2010) Forecasting airplane technologies. Foresight 12(6):38–54
52. Anderson T, Daim T, Lavoie F (2007) Exploratory evaluation of university technology transfer efficiency. Technovation 27(5):306–318
53. Kim J, Daim T, Anderson T (2008) Assessing university technology transfer: measures of efficiency patterns. Int J Innov Technol Manage 5(4):495–526
54. Daim T, Kocaoglu D (2008) Exploring technology acquisition in Oregon, Turkey and in the U.S. electronics manufacturing companies. J High Technol Manage Res 19(1):45–58
55. Basoglu N, Daim T, Topacan U (2012) Determining patient preferences in remote monitoring. J Med Syst 36(3):1389–1401
56. Daim T, Basoglu N, Iskin I, Aydinoglu B, Ozturk M (2012) New product development for the healthcare industry: a case study of diet software. Health Policy Technol 1(2):93–104
57. Ozen C, Daim T, Basoglu N (2010) Exploring user adoption of online services: A prototype driven framework for service design and development. J Enterp Inf Manage 23(4):417–438
58. Kerimoglu O, Basoglu N, Daim T (2008) Organizational adoption of information technologies: Case of enterprise resource planning systems. J High Technol Manage Res 19(1):21–35
59. Basoglu N, Daim T, Kerimoglu O (2007) Organizational adoption of enterprise resource planning systems: A conceptual framework. J High Technol Manage Res 18(1):73–97

# Chapter 2
# No Escape from the Dominant Theories: The Analysis of Intellectual Pillars of Technology Management in Developing Countries

**Berna Beyhan and Dilek Cetindamar**

**Abstract** This chapter aims to identify the intellectual bases of the technology management (TM) literature generated in developing countries using citation and co-citation analyses and answers the question of whether the intellectual bases of the TM literature created by authors in developing countries diverge from those of the global TM literature. Based on a comprehensive bibliometric analysis of ten technology-innovation management (TIM) specialty journals through the period 1998–2007, this study produces three important findings. First, the TM literature generated in developing countries is dominated by the knowledge and theories created in developed countries. Second, among these knowledge sources some authors from developing countries focus on the specialties of developing countries, such as Kim and Lall, come into prominence; however, these authors are not even mentioned in the previous bibliometric studies covering overall TM research. Finally, the researchers in developing countries tackle with the issues or topics specific to their own context through combining three major bulks of literature. These are (1) resource-based view (RBV)/core competencies and organizational learning-related research; (2) literature dealing with the evolutionary theorizing on economic change and growth, and (3) literature related to technological capabilities, technology transfer, and industrialization in developing countries.

Reprinted from Technological Forecasting and Social Change, 78 (1), B. Beyhan and D. Cetindamar, No escape from the dominant theories: the analysis of intellectual pillars of technology management in developing countries, 103–115, (2011), with permission from Elsevier.

B. Beyhan (✉)
Science and Technology Policy Studies Research Center, Middle East Technical University, Inonu Bulvari 06531 Ankara, Turkey
e-mail: berna.beyhan@gmail.com

D. Cetindamar
Faculty of Management, Sabanci University, Tuzla 34956 Istanbul, Turkey

D. Cetindamar et al. (eds.), *Strategic Planning Decisions in the High Tech Industry*, DOI: 10.1007/978-1-4471-4887-6_2,

## 2.1 Introduction

The share of the developing countries in the international technology management (TM) literature has grown rapidly in the recent period; in 2007 nearly one-fourth of the literature was created with the contribution of researchers in developing countries (Table 2.1). However, the TM literature created in developing countries mostly differs from its counterpart generated in the developed world, especially in terms of the research topics it focuses [1, 2]. This study takes a step forward and investigates whether researchers in developing countries utilize the same intellectual pillars with their colleagues in the developed world to understand the issues they specifically focused on. Thus, the chapter provides a comprehensive detailed bibliometric analysis of developing countries' TM literature covering articles published in ten technology-innovation management (TIM) specialty journals [3–6]. Considering that no study employs bibliometric techniques in the field of TM to understand the specific characteristics, knowledge maps, and flows in the literature created in developing countries, this chapter will contribute to the global TM literature. Furthermore, the understanding of theoretical development of TM in developing countries will contribute to the incorporation of particular issues, problems, and theories of developing countries into the TM discipline.

Since the commencement of IEEE Transaction on Engineering Management in 1954 much progress has been accomplished both in the field of management of technology education [7, 8] and in scholarly research with the launch of a number of TIM specialty journals. In this process the number of researches focusing on the scholarly research generated by the scientific community of the TM field has increased. Among those there are a number of studies using bibliometric techniques to evaluate the development of the discipline; however most of them focus on a specific journal [1, 9–12] instead of the whole literature. Furthermore, due to differences between these journals in terms of the issues on which they focus [5] these bibliometric studies cannot provide a clear picture of the field. In spite of the renewed interest in the scholarly communication in TM very few examples [2, 13, 14] focusing on the certain aspects of the literature in developing countries exist. This study aims to overcome these limitations.

The chapter is divided into five sections. Section 2.2 reviews the bibliometric research in TM literature followed by a section on the methodology employed in this chapter. Section 2.4 presents and discusses the results of the empirical study and finally, Sect. 2.5 presents a summary, makes suggestions for future research, and indicates the limitations of the study.

## 2.2 A Review of Bibliometric Research in TM Field

Pritchard [15] provides an early definition of bibliometrics as a method of applying mathematics and statistics to the media of written communication in order to understand the nature and course of development of a discipline. Albeit their

**Table 2.1** Distribution of articles with authors from developing countries

| Journals | 1998 | 1999 | 2000 | 2001 | 2002 | 2003 | 2004 | 2005 | 2006 | 2007 | Total |
|---|---|---|---|---|---|---|---|---|---|---|---|
| Technovation (TVN) | 11 | 13 | 20 | 21 | 21 | 21 | 26 | 36 | 21 | 19 | 209 |
| International Journal of Technology Management (IJTM) | 14 | 9 | 11 | 14 | 12 | 10 | 21 | 29 | 19 | 41 | 180 |
| Technological Forecasting and Social Change (TFSC) | 4 | 13 | 5 | 7 | 7 | 8 | 4 | 17 | 23 | 21 | 109 |
| IEEE Transactions on Engineering Management (IEEE) | 14 | 4 | 4 | 8 | 7 | 7 | 8 | 10 | 10 | 12 | 84 |
| Research Policy (RP) | 5 | 2 | 6 | 9 | 6 | 10 | 7 | 7 | 6 | 9 | 67 |
| R&D Management (RDM) | 3 | 1 | 4 | 7 | 10 | 6 | 7 | 4 | 5 | 9 | 56 |
| Research–Technology Management (RTM) | 1 | 2 | 2 | | 1 | 1 | 2 | 2 | 3 | 4 | 18 |
| Journal of Engineering and Technology Management (JETM) | 1 | 3 | 2 | | 1 | 2 | 1 | | 2 | 3 | 15 |
| Technology Analysis and Strategic Management (TASM) | 2 | 3 | 3 | | | 1 | | 5 | | 1 | 15 |
| Journal of Product Innovation Management (JPIM) | | 1 | | 2 | 2 | 1 | 1 | 1 | 1 | | 9 |
| Total | 55 | 51 | 57 | 68 | 67 | 67 | 77 | 111 | 90 | 119 | 762 |
| % in total TM articles (%) | 14 | 12 | 14 | 16 | 16 | 16 | 17 | 23 | 19 | 24 | 18 |

limitations citation and co-citation analyses are important bibliometric techniques which are widely employed in order to analyze the developments of scientific disciplines or sub-disciplines.

Since the mid-1980s citation and co-citation analyses have been utilized to investigate different facets of management literature and its subfields. Citation and co-citation analyses in management studies have most frequently served for identification of emerging scientific fields/subfields/disciplines, their boundaries, and intellectual developments of these fields [16]. Culnan [17] focused on the intellectual development of the management information systems literature through author co-citation analysis. Pilkington and Liston-Heyes [18] investigated the process in which production and operations management had struggled to establish itself as an academic discipline by examining its literature using citation data obtained from the International Journal of Operations and Production Management. Ramos-Rodriguez and Ruiz-Navarro [19] and Nerur et al. [20] used Strategic Management Journal as a base journal for their citation and co-citation analyses. Among these two complementary studies the first study identified the most influential works on the strategic management research; however, the second delineated the subfields, determined their relationships with each other, and identified the authors bridging two or more conceptual domains of the strategic management research. On the other hand, Martinson et al. [21] focused on journals cited in or cited by Strategic Management Journal and provided a longitudinal portrait of the strategic management sub-discipline and a map of changes in the relationships among journals. Acedo and Casillas [22] used co-citation analysis to understand the intellectual structure of the international management research; Gu [23] focused on the knowledge management literature and identified the authors, journals, research teams in the field; last but not the least Cornelius and Persson [24] provided a bibliometric analysis of the venture capital research.

The TM literature has grown at great speed and become an academic discipline in the last two decades [1, 25] and "once a scientific discipline has reached a certain degree of maturity, it is common practice for its scholars to turn their attention towards the literature generated by the scientific community" [19]. Early examples of studies focusing on the TM field were [25–27]. Adler [25] provided a systematic literature review identifying the underlying themes and concepts related to technology strategy. Drejer [26] dealt with the evolution of the management of technology discipline through four schools of thought which were identified by the author from the literature. On the other hand, Allen and Varghese [27] analyzed the changes in the field of R&D management through articles appearing in R&D Management journal from 1970 to 1987.

In the 2000s, the number of scholars turning their attention toward the TM literature has increased. Among those Beard [28] proposed a model of categorizing the literature on the management of technology; Liao [29] focused on the way in which TM methodologies and applications had developed; and Ball and Rigby [30] investigated the number of author entries and their affiliations throughout 11 journals selected due to their coverage of the R&D and innovation management literature.

On the other hand, there is a bulk of literature analyzing the evolution of certain TIM specialty journals. Linstone [31] reviews the evolution of Technological Forecasting and Social Change from 1969 to 1999. Callon et al. [32] analyze the contents of Research Policy for the first 28 years and state that the journal has succeeded to cover problems of government policy and those of industrial R&D and innovation. For the 50th anniversary of IEEE Transactions on Engineering Management, Allen and Sosa [33] scanned the contents of 50 years and presented a general history of the field of engineering management following the changes in topics, authors' affiliations, and regions throughout five periods. Furthermore, Pilkington [9, 10] reports a bibliometric analysis of IEEE Transactions on Engineering Management. The author uses citation and co-citation analysis, and social network tools to explore the central concepts, theories, and authors of the field and their relationships with each other. Merino et al. [34] focus on the content of Technovation created in its first 25 years. This study uses citation counts to determine the influential journals over the Technovation content and reveals that approximately 68 % of articles drawn from bibliographic references cite TIM journals. Pilkington and Teichert [1] also focus on the TM literature as reported in Technovation. The authors use co-citation and social network analysis techniques to investigate the intellectual structure of the literature. Biemans et al. [12] provide a detailed analysis of Journal of Product Innovation Management (JPIM)'s evolution from 1984 to 2003. The authors investigate the contents of the journal to identify main research areas, research methods used in the articles, authors' background and affiliations; and finally the knowledge sources used by JPIM authors while developing their articles with using bibliometric analysis. McMillan [11] employs citation and co-citation analyses to examine R&D Management throughout four periods and aims to reveal changes in the journal's intellectual base.

Among these studies, [12, 27, 32, 34] partially utilize citation analysis. Callon et al. [32] explore the most cited articles among those published in Research Policy; Merino et al. [34] use citation analysis to determine the influential journals over the Technovation content; and Biemans et al. [12] identify journals which are most frequently cited in and citing JPIM. Pilkington and Teichert [1], Pilkington [9, 10] and McMillan [11] provide good examples of citation and co-citation analyses employed to fully understand and map out the intellectual pillars as well as the invisible colleges of the discipline. Moreover, two additional studies [3, 35] which utilize citation analysis to determine the most influential journals of TM must be mentioned.

None of these aforementioned studies focusing on the TM literature have paid special attention to the contribution of developing countries to the literature and the particularities of this contribution especially in terms of themes, theories, knowledge sources, and flows. However, three studies need to be mentioned as the very occasional examples of studies investigating the particularities of the TM literature generated in developing countries; these are [2, 13, 14]. Seol and Park [14] analyze the knowledge sources of Korean innovation studies using citation analysis; they identify the most highly cited papers, books, authors, and journals in academic studies carried out by Korean researchers. Some studies [32–34] prove that the majority of papers published in the prominent TIM-specialty journals are submitted by authors affiliated to institutions in North America and Western Europe. However, among those studies only [1] have questioned whether authors from geographically different regions exhibit different citation patterns. The results of citation and co-citation analyses indicate significant differences in the intellectual interests of authors from different regions which are categorized into four as North America, Europe, UK, and rest of the world. Cetindamar et al. [2] state that "there are substantial differences among the topics investigated in developed and developing country studies". Following the footsteps of previous studies (such as [1, 2, 14]) which investigate the particularities of TM research generated in the different parts of the world, this study examines, how the scholars in developing countries benefit from the previous literature in dealing with the topics which are proved to be different from those in the developed countries [1, 2]. Therefore, in this study we not only consider the most highly cited academic studies, authors, or journals by the scholars in these countries in order to support their views, ideas or methods, but also how they tie these different pieces of literature in their researches or articles. This study contributes to the aforementioned literature investigating the TM literature itself by mapping the intellectual sources of TM literature produced by developing country scholars.

## 2.3 Data Collection and Methodology

The data used in this study include authors, keywords, the name and addresses of institutes, publication dates, source titles, and references of ten leading TIM specialty journals (JPIM, Research Policy, Research-Technology Management, R&D Management, IEEE Transactions on Engineering Management, Technological Forecasting and Social Change, International Journal of Technology Management, Technovation, Technology Analysis and Strategic Management, and the Journal of Engineering Technology Management) for a 10 year period, 1998–2007. The identification of leading journals in the TM field is important but difficult because major journals of the discipline are not as apparent as those in established academic fields due to the interdisciplinary character of the field [3, 35]. There are some studies [3, 30, 35–37] that attempted to identify the major journals in the TM field and rank them according to their importance to the field. Some of these studies are based on the subjective opinions of scholars [30, 36] and some are on citation-based analysis [3, 35]. The aforementioned ten journals we are using in this study are those identified by Linton and Thongpapanl [3]. Since, first of all at least four different studies [3, 30, 35, 37] with various methodologies identify all or most of these ten journals of the TM field and second in spite of the years elapsed after the publication of [3] the same journals are still considered as the specialty journals in the management of technology and innovation [4–6], we prefer to rely on this list of journals in our study.

Distinguishing between developed and developing countries is the key to this research. However, there is no established convention on the distinction between developed and developing countries or areas. Moreover, for the purpose of this study, any list of country classifications based on income level or the level of human development is not very useful by itself since such lists do not provide evidences for the countries' technology and innovation management capabilities. Therefore, inclusion of some countries which are not listed as a developing country by some global institutes, i.e., World Bank, OECD, or IMF in our research is simply based on this consideration. Instead of making a list of developing countries we first decided on the countries which are developed not only in the sense of productivity, industrialization, or income but also in the sense of technology and innovation management skills and capabilities of their firms. To start we used the list of 25 high-income OECD countries as of 25 July 2008 [38]; however, we excluded the Czech Republic and South Korea in our list of developed countries because TM practices and experiences in these two countries are "more closely related to circumstances in developing countries rather than developed countries" [2].

The bibliometric data including the full contents of these journals for a 10-year period 1998–2007 were collected from the ISI Web of Science databases on 16 January 2009 (for the query see Appendix A). After the first set of results had been received, using pull-down menu on the Web page the results were further refined to include only the original articles written by at least one author located in developing countries. We first removed book reviews, editorials, and brief notes from the set of results and we were left with 4,349 original articles. In order to identify the articles with a developing country author or co-author a similar refinement process was repeated. We visualize a complete list of 74 countries (Appendix B) contributed to the articles by using pull-down menu, tagged only those countries out of 23 high-income OECD countries we selected, and refined the results based on this selection. Finally we had a list of 764 articles in which at least one researcher was linked to a developing country institution. Full bibliometric records (including cited references) of these articles were exported as a text file from the ISI Web of Science. During the detailed examination of the data file, two more articles were discarded because only their reprint addresses include a developing country institute but authors were affiliated to developed country institutions. The number of these 762 articles by publication year and journal is shown in Table 2.1.

The records of 762 articles retrieved from the ISI Web of Science were reformatted into a Microsoft Access 2003 database using a Visual Basic script. Each of these articles was given a unique number from 1 to 762 and all variables included in bibliometric content (i.e. authors, addresses, titles, keywords, and references) are linked to each other through this unique identifier. Data manipulation and analyses were performed through created tables and queries in this database. Most of these tables and queries were recreated from bibliometric software tool Sitkis [39] which is also based on Microsoft Access. These different tables are used simple counting of articles, keywords, or citations and queries allow matching different tables by the unique identifier in order to count the frequency of simultaneous occurrences of two different elements (i.e. citations and keywords) in the same document. This tool also allowed the manipulated data to be exported to MS Excel and UCINET [40] compatible tables. The networks were analyzed using the social network analysis software UCINET and were drawn with NetDraw package embedded to UCINET.

The developed database was analyzed in order to identify the intellectual pillars of the developing country TM literature through the most cited academic studies, authors, and journals and the invisible colleges through examining how different pieces of previous academic studies and their authors had been linked in these set of articles. For this aim, citation and co-citation analyses were used. Citations are widely used tools for understanding the linkages between academic studies; the exchanges among scholars and hence scholarly knowledge flows [17, 41]. Citation analysis is based on the argument that authors cite papers which embody the ideas they are discussing; and therefore these cited documents, in a general sense, are the symbols for these ideas [42]. On the other hand, co-citation analysis measures the frequency with which two documents are cited together [43]. It is widely assumed that co-citation patterns delineate the relationships between key ideas and therefore provides an objective way of modeling not only the intellectual structure of scientific fields/subfields but also their historical developments [43, 44]. Hence, citation and co-citation analyses provide a well-established procedure for measuring the dissemination and the extent of knowledge exchange in a given field [35], social and cognitive structure of research specialties [45] and identification of 'invisible colleges' [11, 46, 47] which focus on common problems in common ways.

For the citation analysis the cited references in 762 articles were first collected in a table. However 1,926 citations were immediately discarded because of their improper formats; those also include newspaper or magazine articles, various reports, or documents. We started with 18,558 citations derived from 762 articles; however, 695 of them were also removed to correct the problem of multiple entries occurred due to the inconsistencies in the spelling of author names, journals, and volume or page numbers. After all, citation and co-citation analyses are carried out with these 17,863 documents cited in these articles. In order to find out the most cited academic studies, journals, and authors we simply counted how many times an academic study, a journal name, or an author name had occurred in our data set. Co-citation analysis was also carried out through the counts of the co-occurrences of two different citations in the same document. However, in co-citation analysis of academic studies and authors we counted the number of articles in which two academic studies or the names of two authors co-occurred. The tables including the data regarding to co-citation networks are imported into UCINET [40] and the standard centrality measures of degree, closeness, and betweenness are calculated. Netdraw is utilized for the visualization of networks. The same procedure was also repeated for the analysis and visualization of the keywords co-occurred in our set of articles.

As a final point, in these visual representations of networks the thickness of lines between nodes reflects the strength of the link which is measured by counts of the frequency with which the two items co-occurred and the size of circles indicates the degree centrality of nodes in the network; the higher the degree of centrality the higher the size of circles. Degree centrality "measures the extent to which a node connects to all other nodes in a social network" [48]. In network studies it is proposed that nodes or actors with higher number of ties with other nodes may have advantaged positions since they occupy a more central position than those having lower number of ties.

## 2.4 Results

The TM literature has grown at an exponential pace especially in the last few decades. The number of articles published in these TIM specialty journals increased by nearly 164 % from the period of 1986–1994 to 1995–2005 and in the same period the number of articles focusing on developing countries increased by 388.5 % [2]. On the other hand, the number of articles having at least one author affiliated to an institution in developing countries has displayed a 116 % increase in the period 1998–2007. The share of the developing countries in the international TM literature has grown rapidly in this period; in 2007 nearly one-fourth of the literature was created by the contribution of researchers in developing countries (Table 2.1).

The initial analysis reveals that 762 articles in our dataset are produced by 1,237 authors from 66 countries of which 15 are classified as developed countries in this study. This discrepancy is due to the fact that some authors affiliated to developed country institutions are the co-authors of 177 articles which account for 91 % of the whole international joint publications in the dataset. Moreover, due to international co-authorships some articles are counted more than once. Taiwan is the most productive country with 180 articles published in these ten TIM specialty journals between the years 1998 and 2007 and the total citations of the articles created by researchers affiliated to Taiwanese institutions is the highest among all other countries. On the other hand Brazil, Singapore, Israel, China, and South Korea produce higher number of articles with better citation per article ratios (Table 2.2). The US, UK, Australia, and Netherlands occupy higher ranks in Table 2.2 as the most collaborating developed countries with the so-called developing ones in the field of TM. On the other hand while the number of internationally collaborated articles increases by years its share in the whole number of articles changes in a range from 17.6 to 32.2 %. Among the developing countries China and South Korea are the first and the second countries, respectively, which have the highest number of international links.

In order to scrutinize the main issues and topics discussed in these 762 articles we carried out a simple keyword co-occurrence analysis and mapped out how the keywords selected to define the content of the articles are linked to each other. Figure 2.1 presents a network representation of the most frequently occurred keyword couples in the articles and shows those words which appear together in the same document more than seven times. The graphical representation of keyword co-occurrence network provides some clues about the main issues on which developing countries focused and about where developing countries' contribution to the international TM literature is concentrated.

**Table 2.2** Distribution of articles by country of origin (first 15 countries)

| Rank | Country | Number of articles | Total citation | TC/number of articles | Rank acc. to sum of citations |
|---|---|---|---|---|---|
| 1 | Taiwan | 180 | 575 | 3.19 | 2 |
| 2 | USA | 103 | 684 | 6.64 | 1 |
| 3 | South Korea | 103 | 439 | 4.26 | 4 |
| 4 | People's Republic of China | 100 | 457 | 4.57 | 3 |
| 5 | India | 75 | 252 | 3.36 | 6 |
| 6 | Singapore | 52 | 316 | 6.08 | 5 |
| 7 | Israel | 40 | 200 | 5.00 | 9 |
| 8 | UK | 34 | 245 | 7.21 | 7 |
| 9 | Brazil | 34 | 218 | 6.41 | 8 |
| 10 | Nigeria | 33 | 53 | 1.61 | 16 |
| 11 | Turkey | 21 | 74 | 3.52 | 14 |
| 12 | Thailand | 21 | 44 | 2.10 | 17 |
| 13 | South Africa | 18 | 69 | 3.83 | 15 |
| 14 | Australia | 13 | 42 | 3.23 | 18 |
| 15 | Netherlands | 12 | 79 | 6.58 | 12 |

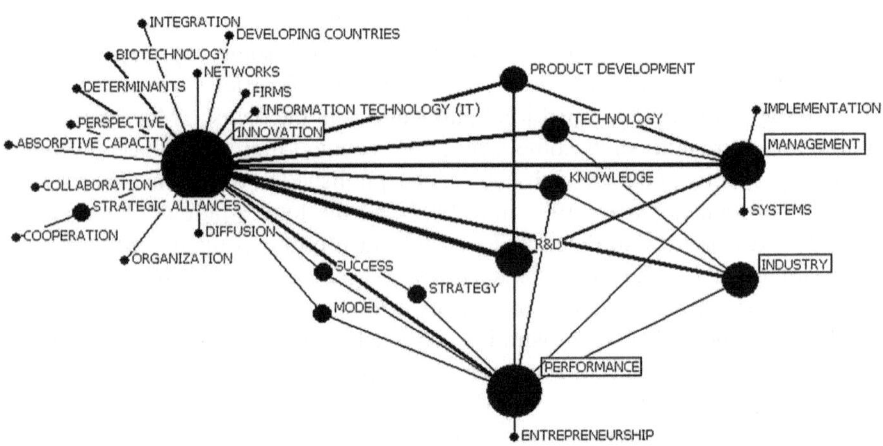


**Fig. 2.1** Keyword co-occurrence network

Figure 2.1 helps identify the focus of the articles created by developing country authors. While "innovation" is at the center of the network, some keywords such as "performance", "management", "R&D", and "industry" are the other central terms in the map. By inspection we can identify some topics that the TM literature in developing countries is focused on such as *innovation management, R&D management, product development through R&D, networks and collaborations for innovation, emerging technologies (biotechnology and information technology) and innovation, determinants of innovation, diffusion of innovations, absorptive capacity, organizational change and innovation, R&D performance, innovation performance, industrial innovation, TM, innovation strategy, innovation in developing countries* etc.

A cursive glance at the map implies that the relationships between different areas of TM interests in developing countries do not diverge dramatically from general themes of the global TM literature. However, in the rest of this study we are going to further scrutinize the distinguishing interests of developing country authors and how they approach their subjects of interests by understanding the intellectual structure of TM studies through citation and co-citation analysis of the articles in our sample.

### 2.4.1 Citation Analysis: Intellectual Pillars of TM Literature in Developing Countries

Citation analysis is generally used to detect the most influential documents or authors over the later ones. Our analysis of the main knowledge sources of the developing country TM literature uses the cited references in 762 articles in our dataset in order to understand the most influential sources of the literature at three levels (1) articles/books, (2) authors, and (3) journals.

Table 2.3 gives a list of the most frequently cited individual academic studies. Most of these studies in the list question the role of strategy, differences in technology policy or performance, and strategy by focusing at the level of nation and industry. These findings are consistent with [1] providing that academic studies which focus on national systems, diffusion, and adoption of innovations are cited heavily by the authors out of North America, Europe, and the UK. These findings also provide additional confirming evidence for [2] showing that technology policy-related issues, i.e., national TM policies and systems, innovation systems, national innovation systems, regional innovation systems, sectoral innovation systems, open innovation system are the most studied topics in developing country articles. However, this special focus on national systems and differences does not fully eliminate the academic studies concentrating on the firm level analysis of competitiveness and the sources of competition. The higher ranks of four studies in Table 2.3 [49–52] prove the interests of these authors toward the issues related to resource-based view (RBV)/core competencies and organizational learning. Furthermore, the first rank is occupied by Cohen and Levinthal's study [49] on "absorptive capacity". The same study is listed at the second rank in [1], and at the first rank in [11] among the most cited articles in the period 2001–2005. The glaring presence of Kim's book "Imitation to Innovation" [53] in the sixth rank in Table 2.3 provides a strong support for the arguments about the particularities of the technological development and the importance of technology adoption and learning in developing countries.

Table 2.4 shows the frequencies of authors of the cited documents. Although this list is based merely on the first authors and includes bias against younger

**Table 2.3** Most frequently cited academic studies

| First authors | Title (volume) | Years | Number of citations | Number in references |
|---|---|---|---|---|
| Cohen WM | Administrative Science Quarterly (35) | 1990 | 50 | [49] |
| Nelson RR | National Innovation Systems: A Comparative Analysis | 1993 | 49 | [64] |
| Porter ME | The Competitive Advantage of Nations | 1990 | 46 | [56] |
| Nelson RR | Evolutionary Theory of Economic Change | 1982 | 45 | [50] |
| Nonaka I | The Knowledge Creating Company: How Japanese Companies Create The Dynamics of Innovation | 1995 | 39 | [62] |
| Kim L | Imitation to Innovation: The Dynamics of Korea's Technological Learning | 1997 | 36 | [53] |
| Freeman C | The Economics of Industrial Innovation | 1974[a] | 31 | [81] |
| Lundvall BA | National Systems of Innovation : Towards a Theory of Innovation and Interactive Learning | 1992 | 31 | [67] |
| Rogers EM | Diffusion of Innovations | 1962[a] | 31 | [82] |
| Dosi G | Research Policy (11) | 1982 | 27 | [69] |
| Porter ME | Competitive Advantage: Creating and Sustaining Superior Performance | 1985 | 27 | [55] |
| Leonard-Barton D | Wellsprings of Knowledge: Building and Sustaining The Sources of Innovation | 1995[a] | 26 | [59] |
| Barney J | Journal of Management (17) | 1991 | 24 | [51] |
| Porter ME | Competitive Strategy: Techniques for Analyzing Industries and Competitors | 1980 | 23 | [54] |
| Prahalad CK | Harvard Business Review (68) | 1990 | 23 | [52] |
| Utterback JM | Mastering The Dynamics of Innovation: How Companies Can Seize Opportunities in The Face of Technological Change | 1994 | 23 | [83] |
| Utterback JM | Omega-International Journal of Management Science (3) | 1975 | 23 | [84] |
| Eisenhardt KM | Academy of Management Review (14) | 1989 | 21 | [85] |
| Teece DJ | Research Policy (15) | 1986 | 21 | [86] |

[a] Citations to later editions of the books were also included

**Table 2.4** Most frequently cited first listed authors

| Rank | Authors | Number of citations |
|---|---|---|
| 1 | Porter ME | 152 |
| 2 | Nelson RR | 132 |
| 3 | Kim L | 109 |
| 4 | Freeman C | 104 |
| 5 | Cohen WM | 96 |
| 6 | Cooper RG | 94 |
| 7 | Teece DJ | 84 |
| 8 | Nonaka I | 76 |
| 9 | Griliches Z | 75 |
| 10 | Lall S | 72 |
| 11 | Dosi G | 71 |
| 12 | Pavitt K | 69 |
| 13 | Utterback JM | 65 |
| 14 | Rothwell R | 63 |
| 15 | Leonard-Barton D | 63 |

authors [1] it provides some insights into some authors' strong influences on the discipline. Among those influential authors Porter, Nelson, Kim, and Freeman are prominent with more than a hundred citations. Porter is the most cited author thanks to his famous three books [54–56]. In the list, Kim in the third rank and Lall in the tenth rank are from developing countries.

The list of most cited journals is given in Table 2.5. This list indicates the problem stated by Pilkington and Teichert [1] regarding the TM's having become an academic discipline. Similar to their list, the general management and strategy specific journals occupy highly prominent places in the list of the most cited journals. While a journal related to the development issue, World Development is listed as among the 30 most cited journals, operations management focused titles are lower ranked in comparison to the list in [1]. The comparison of the rankings of journals in these two lists (Table 2.5 in this study and Table 2.2 in [1]) with 'Wilcoxon signed rank test' points out the fact that the rankings of these journals are significantly ($p \leq 0.05$) different; the paired samples $t$ test also produces the same result. This may indicate that knowledge sources of the TM literature in developing countries differ from those of the international TM literature at the journal level.

Table 2.5 supports the findings of previous studies [1, 3, 35] which emphasize the great variety of cited journals from different disciplines and sub-disciplines of management, social sciences, economics as well as those in science, engineering, and technology. While the number of articles has increased by 116 % the number of citations has increased by 227 % from 1998 to 2007. This increase in the number of references in articles indicates a growth in the developing country TM literature. On the other hand citations of TIM-specialty journals have increased by 123.3 % from the first period of 1998–2002 to the second period; its share in total has stayed nearly the same. Cheng et al. [35] have found that as citation data is considered TM discipline "has not shown any symptoms of inbreeding" and the

**Table 2.5** Most frequently cited journals (ranked based on the number of citations in the period 2003–2007)

| Journals | 1998–2002 | 2003–2007 | Increase (%) |
|---|---|---|---|
| Research Policy | 193 | 607 | 214.5 |
| Strategic Management Journal | 114 | 453 | 297.4 |
| Management Science | 138 | 261 | 89.1 |
| Harvard Business Review | 97 | 256 | 163.9 |
| Technovation | 78 | 255 | 226.9 |
| Journal of Product Innovation Management | 106 | 250 | 135.8 |
| R&D Management | 106 | 235 | 121.7 |
| Academy of Management Journal | 52 | 223 | 328.8 |
| Administrative Science Quarterly | 65 | 210 | 223.1 |
| Organization Science | 30 | 197 | 556.7 |
| Academy of Management Review | 62 | 192 | 209.7 |
| IEEE Transactions on Engineering Management | 143 | 188 | 31.5 |
| Technological Forecasting and Social Change | 81 | 182 | 124.7 |
| Journal of Marketing | 65 | 174 | 167.7 |
| Research (Technology) Management | 106 | 172 | 62.3 |
| International Journal of Technology Management | 101 | 159 | 57.4 |
| Sloan Management Review | 67 | 117 | 74.6 |
| Journal of Marketing Research | 38 | 113 | 197.4 |
| California Management Review | 60 | 102 | 70.0 |
| American Economic Review | 20 | 100 | 400.0 |
| MIS Quarterly | 16 | 97 | 506.3 |
| Journal of Business Venturing | 21 | 85 | 304.8 |
| European Journal of Operations Research | 21 | 83 | 295.2 |
| Industrial Corporate Change | 17 | 81 | 376.5 |
| Journal of Management | 16 | 80 | 400.0 |
| Journal of International Business Studies | 17 | 72 | 323.5 |
| Long Range Planning | 33 | 71 | 115.2 |
| Journal of Operations Management | 20 | 67 | 235.0 |
| Information and Management | 30 | 63 | 110.0 |
| Journal of Engineering and Technology Management | 16 | 63 | 293.8 |
| Journal of Political Economy | 15 | 60 | 300.0 |
| World Development | 51 | 55 | 7.8 |
| OMEGA-International Journal of Management Studies | 23 | 54 | 134.8 |
| Technology Analysis and Strategic Management | 33 | 39 | 18.2 |
| Total of Ten TIM Journals | 963 | 2150 | 123.3 |
| Total | 7,968 | 16 232 | 103.7 |
| TIM Journals/Total (%) | 12.1 | 13.2 | |

authors contributing to the discipline are generally more open to citing from sources outside of TM. However, this great variety in citations can also be explained by the fact that TM researchers prefer to publish their work in more established journals such as Administrative Science Quarterly, Management Science, The Academy of Management Journal, Harvard Business Review, or Strategic Management Journal [1].

While the above lists of most cited documents, authors, and journals provide some insight into the intellectual pillars of the TM literature created by developing country scholars, they are not sufficient to have a clear picture of how different academic studies and authors are linked to each other and hence where the interests of these scholars are concentrated. The answers to these questions are provided through co-citation analysis [43–45] and the main findings are presented in the following section.

### 2.4.2 Co-Citation Analysis: Invisible Colleges

The network for co-cited academic studies which appear together with a frequency greater than seven is shown in Fig. 2.2. Three segments of literature as grounding the TM literature generated in developing countries can be easily identified in the map. One of these segments is centered on the seminal work of Cohen and Levinthal [49] and includes academic studies mostly related to RBV/core competencies and organizational learning, i.e. [57–62]. RBV is based on the argument that "sources of sustained competitive advantage are firm resources that are valuable, rare, imperfectly imitable, and non-substitutable" [51]. Organizational knowledge is seen as the most strategically significant resource of the firm [63].

The second segment of literature in the map is focused on national systems, centered on the influential book edited by Nelson "National innovation systems: a comparative analysis" [64]. Heterogeneity of economic agents operating in national innovation systems [50] is the starting point in the evolutionary theorizing for the understanding of complexities associated with the process of growth and transformation in the long run [65, 66]. A systemic approach [64, 67, 68] provides a framework to investigate these complexities. Another strand of the literature contributing evolutionary growth theorizing covers some academic studies focusing on the macro economic impacts of radical innovations in the long run [69, 70]. In this segment, a very strong link exists between [64] and [67] both focused on national innovation systems; these two studies appear together in 17 of the articles.

The third segment of the literature is centered on the seminal work of Korean author Kim [53]. Studies in this group attempt to understand differences in technology policy and performance. They include the articles focusing on the particularities of developing countries, especially those of Asian newly industrialized countries, in terms of the development of technological capabilities, technological accumulation and change, industrialization, and growth [71–76]. The position of [49] as bridging different segments of the literature needs further discussion. Zahra and George [77] highlight that Kim's [78] definition of "absorptive capacity" requires learning capability which is the capacity to assimilate knowledge for imitation and develops problem solving skills to create new knowledge for innovation. Kim [79] argues that technological trajectory has been reversed in developing countries; starts with mature technology state (for duplicative imitation), evolves to the intermediate technology stage (for creative imitation), and

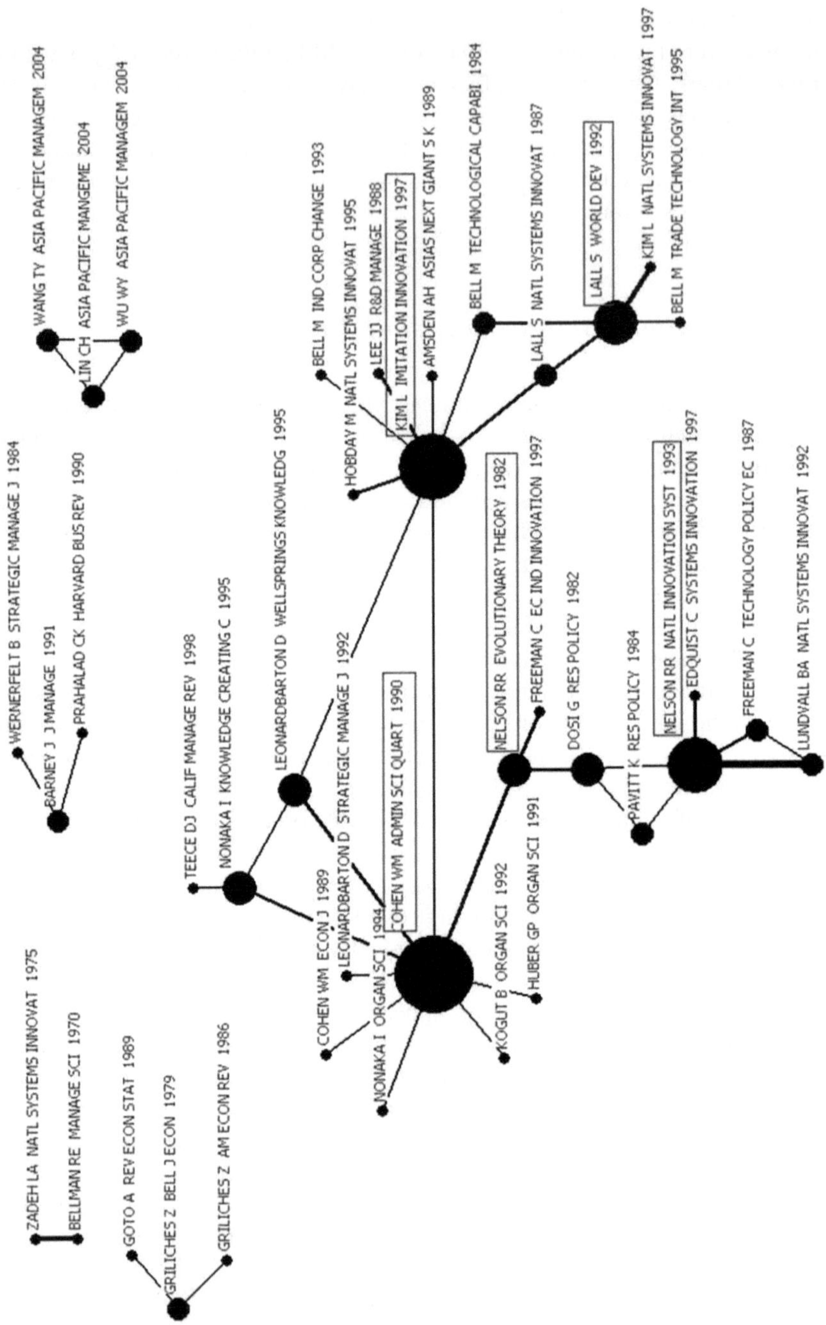


**Fig. 2.2** Core literature-network of co-cited articles/books

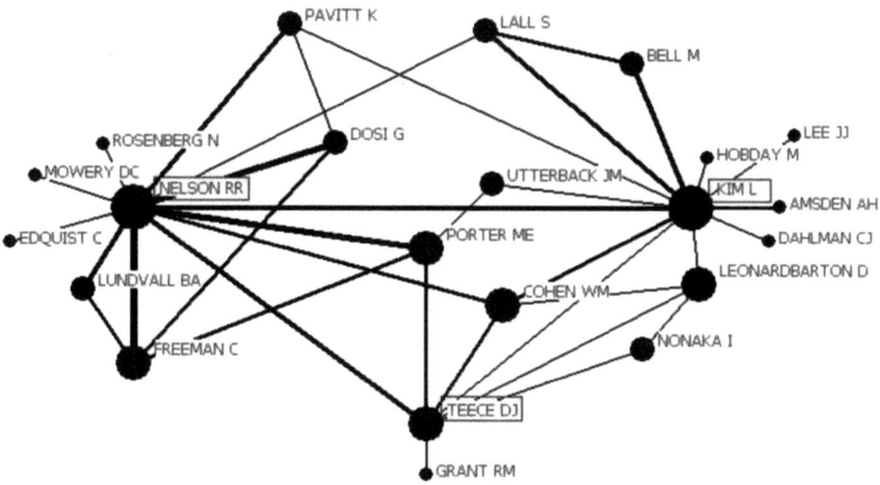


**Fig. 2.3** Network of co-cited authors

finally to the emerging technology stage (for innovation). In this process, from imitation to innovation technological capabilities acquired through learning play a very crucial role; and "effective technological learning requires absorptive capacity" [53]. The emphasized role of technological learning and absorptive capacity in the development of technological capabilities, technological change, and industrialization explains why [49] is so heavily cited by authors affiliated in developing countries and why it is closely linked to Kim's famous book "Imitation to Innovation". Hence, the number one rank occupied by [49] in Table 2.3 and its central position in the cognitive map of TM literature in developing countries graphically represented in Fig. 2.2 indicate that the concept of "absorptive capacity" is widely used by authors from developing countries; however, it has been rebuilt or reified as different from the original ones and from its counterparts rebuilt by the authors of the other regions.

Author co-citation analysis provides insights into underlying schools of thought in scientific discourse [44]. The underlying arguments of author co-citation analysis are that studies of the same author represent a body of knowledge and authors having related works are cited together. Figure 2.3 shows a representation of the network for co-cited authors which appear together in at least 20 of the articles in our dataset.

Nelson, Kim, and Teece seem to be at the center of the co-citation map with their higher number of links to other authors. The body of knowledge created by Nelson is not only linked to Neo-Schumpeterian view such as Freeman, Dosi, and Pavitt focusing on technological change and industrialization or to strategy-related topics (Porter and Teece), but also to the literature on industrialization in developing countries represented by Kim and adjacent authors. Kim also occupies a central position in the map of knowledge sources of the literature generated in developing countries.

Studies focusing on single TIM-specialty journals [9–11] include mostly different schemes of co-cited authors. The central positions of Kim and the segment of literature including Kim, Lall, Lee, Amsden, and Bell present the main particularities of the TM research generated with the contribution of developing country authors. The networks of co-cited authors and co-cited documents reveal that although researchers in developing countries utilize similar knowledge pillars and intellectual bases of the global TM literature, in order to understand and analyze some issues specific to the context of developing countries such as industrialization, technology adoption, or transfer they use a different body of knowledge focusing on the specialties of technology development and industrialization in developing countries and partly created by scholars in developing countries such as Kim, Lall, and Lee.

In the process of the emergence of TM as a discipline in its own right the major academic works, in other words "discipline forming titles" [1], and invisible colleges have been substantially occurred. Undoubtedly, along this dynamic process some academic studies are replaced by others [11]. Yet we have a more or less formed map of TM knowledge areas [1, 9–11] which are associated with certain authors and academic studies. The analyses carried out in this research show that although the TM literature created by the scholars in the developing countries can be distinguished from its counterparts in terms of the issues questioned in the articles, the knowledge bases grounding their research mostly covers the academic studies or authors which are globally associated with these issues i.e. [49, 50, 57–59, 64, 67–70]. The only particularity in the map of literature grounding TM literature in developing countries occurs in the issues related to the development of technological capabilities, technological change, and industrialization in developing countries.

## 2.5 Concluding Remarks, Future Research, and Limitations

This study mainly investigates the intellectual structure of the academic TM research in developing countries. The contribution of developing countries to the international TM literature has been growing substantially in the last decade [2]. Our analysis shows that in 2007 nearly one-fourth of the articles published in the ten specialized journals in TIM under study are written with the contribution of at least one author affiliated to developing country institutions. Although [1, 2] provide evidence that the TM literature generated in developing countries differs from its counterparts generated in developed countries in terms of themes and concepts that are focused on, these studies do not present a detailed analysis of knowledge sources and intellectual pillars that developing country studies are based on. Nonetheless, the understanding of theoretical development of TM in developing countries will also contribute to the incorporation of particular issues, problems, and theories of developing countries into the TM discipline.

Based on a comprehensive bibliometric analysis, this study produces three important findings. First, the TM literature generated in developing countries is dominated by the knowledge and theories created in developed countries. The most influential studies and authors on the TM research held in such countries underline that developing country researchers utilize theories that are created in developed countries to understand even the issues specific to developing countries. Not only studies published in international journals but Seol and Park [14] show that those published in national journals also mostly utilize foreign knowledge sources. Second, among these knowledge sources, some authors from developing countries such as Kim and Lall and authors with a focus on the specialties of developing countries like Bell and Amsden, come into prominence; however these authors are not even mentioned in previous bibliometric studies covering overall TM research. Finally, although the TM literature created in developing countries is mainly dominated by the knowledge and theories coming from developed countries, researchers in such countries tackle with the issues or topics specific to their own context through combining three major bulks of the literature. These are (1) RBV/core competencies and organizational learning-related research; (2) literature dealing with the evolutionary theorizing about economic change and growth, and (3) literature related to technological capabilities, technology transfer, and industrialization in developing countries.

In [1], the results of citation and co-citation analyses indicate significant differences in the intellectual interests of authors from different regions which are categorized into four as North America, Europe, UK, and rest of the world. Cetindamar et al. [2] also emphasize the substantial differences among the topics investigated in developed and developing country studies. Our study provides further evidence that developing country researchers utilize different knowledge sources to understand these divergent topics they deal with. The authors, studies, or journals they mostly cite exhibit certain differences from those cited by their colleagues affiliated to developed country institutes. However, while doing so they utilize the seminar works of the TM literature and combine these theories with a strand of the literature on the specificities of industrialization in developing countries which is partly created by the authors of developing countries. As we consider citations as "the symbols of concepts or methods" [42] this study shows that the symbols used by TM scholars in developing countries such as [49, 50, 54–56, 59, 62, 64, 69] do not fully diverge from those of the global TM literature as listed by Pilkington and Teichert [1].

One likely avenue for future research is the investigation of the observed convergence of theories. Our study shows that developing country researchers utilize theories created in developed countries to understand even the issues specific to developing countries; however, our study does not pass any judgment on reasons/mechanisms behind it. One likely explanation for convergence of theories might be the argument of the diffusion of capitalist economical and industrial infrastructures in developing and developed countries. Another, rather skeptic, explanation might be the limitation faced by authors in developing countries when they submit their papers to theoretical outlets such as management journals might not accept out-of-the-norm papers in the TM field, as is the case for management studies [80]. Another interesting stream of research could be the analysis of international co-authorship patterns. As the number of studies co-authored by developed and developing country scholars is increasing, their analysis might add new insights into the process of theory building in TM in different parts of the world.

This study is rather limited to report on what is being studied, not what should be studied for developing countries. We believe what should be studied that is not part of the developed country model can offer interesting findings for developing country researchers. For example, appropriate technology and technology integration are two such areas that can shape the research projects with high value of practical implications. We strongly advise TM scholars to conduct research in this avenue of research.

Finally, we have to mention that this study inevitably has its limitations, some resulting from the research design and some from the bibliometric techniques we use for analyses. The main limitation related to the research design arises from the selection of journals as TM outlets. As mentioned by Pilkington and Teichert [1] and Cheng et al. [35] the researchers in the TM field prefer to publish their works in more established management journals rather than TIM-specialty journals. Indeed, some significant changes might occur in these analyses if the range of journals included were extended. However, when the number of articles included in the research is considered it can be confidently argued that the literature analyzed in this study represents the major efforts of developing countries researchers in the TM field. Limitations as direct consequences of bibliometric studies are mainly due to the fact that citation and co-citation analyses are independent from the context, or in other words it is not possible to distinguish the motives behind these citations [19]. However, in spite of all these limitations bibliometric analyses employed in this study provide a good insight into the development of the TM discipline created with the contribution of developing country researchers; and how it diverges from its counterparts generated in the developed world especially in terms of knowledge sources employed.

## Appendix A

The initial query used in order to find out the articles published in ten TIM-specialty journals between the years 1998 and 2007.

Publication name (SO) = RESEARCH POLICY OR TECHNOVATION OR R&D MANAGEMENT OR RESEARCH TECHNOLOGY.

MANAGEMENT OR IEEE TRANSACTIONS ON ENGINEERING MANAGEMENT OR JOURNAL OF ENGINEERING AND TECHNOLOGY.

MANAGEMENT OR INTERNATIONAL JOURNAL OF TECHNOLOGY MANAGEMENT OR JOURNAL OF PRODUCT INNOVATION.

MANAGEMENT OR TECHNOLOGICAL FORECASTING AND SOCIAL CHANGE OR TECHNOLOGY ANALYSIS AND STRATEGIC MANAGEMENT.

Publication years (PY) = 1998–2007.

Databases = SCI-EXPANDED, SSCI, A&HCI.

# Appendix B

(Table 2.6).

**Table 2.6** Distribution of 4,349 original articles published between 1998 and 2007 by country of origin

| Developed countries | Number of articles | Developing countries | Number of articles | Developing countries | Number of articles |
|---|---|---|---|---|---|
| Australia | 131 | Argentina | 5 | Lebanon | 1 |
| Austria | 64 | Bahrain | 1 | Lithuania | 1 |
| Belgium | 48 | Bangladesh | 1 | Malaysia | 5 |
| Canada | 184 | Barbados | 1 | Mexico | 11 |
| Denmark | 82 | Bolivia | 1 | Moldova | 1 |
| Finland | 75 | Botswana | 1 | Morocco | 1 |
| France | 166 | Brazil | 34 | Nigeria | 33 |
| Germany | 213 | Bulgaria | 1 | Oman | 1 |
| Greece | 33 | Chile | 5 | People's Republic of China | 100 |
| Iceland | 1 | Colombia | 1 | Poland | 7 |
| Ireland | 23 | Croatia | 2 | Romania | 2 |
| Italy | 171 | Cuba | 1 | Russia | 4 |
| Japan | 158 | Cyprus | 8 | Saudi Arabia | 6 |
| Luxembourg | 1 | Ecuador | 1 | Singapore | 52 |
| Netherlands | 294 | Egypt | 3 | Slovenia | 4 |
| New Zealand | 28 | Fiji | 1 | South Africa | 18 |
| Norway | 21 | Ghana | 1 | South Korea | 103 |
| Portugal | 33 | Hong Kong | 7 | Taiwan | 181[b] |
| Spain | 135 | Hungary | 6 | Thailand | 21 |
| Sweden | 127 | India | 75 | Trinidad and Tobago | 3 |
| Switzerland | 80 | Indonesia | 3 | Turkey | 22[b] |
| UK (England) | 571 | Iran | 1 | U. A. E. | 2 |
| UK (N. Ireland) | 13 | Israel | 40 | Uganda | 1 |
| UK (Scotland) | 65 | Jordan | 2 | Uruguay | 2 |
| UK (Wales) | 28 | Kenya | 2 | | |
| USA | 1,496 | Kuwait | 3 | | |
| San Marino[a] | 2 | | | | |

[a] Although San Marino was not among our list of developed countries, the articles co-authored by scholars from San Marino were excluded from the set of results because it is a very small country completely surrounded by Italy and both these articles are about Italy

[b] One article was discarded because only its reprint address includes an institute from this country

## References

1. Pilkington A, Teichert T (2006) Management of technology themes, concepts and relationships. Technovation 26:288–299
2. Cetindamar D, Wasti SN, Ansal H, Beyhan B (2009) Does technology management research diverge or converge in developing and developed countries. Technovation 29:45–58
3. Linton JD, Thongpapanl N (2004) Ranking the technology innovation management journals. J Prod Innov Manag 21:123–139
4. Linton JD (2006) Ranking of technology innovation management journals. Technovation 26:285–287
5. Linton JD, Embrechts M (2007) MOT TIM journal rankings 2006. Technovation 27:91–94
6. Linton JD (2009) Technology innovation management's growing influence and impact. Technovation 29:643–644
7. Reisman A (1994) Technology management: a brief review of the last 40 years and some thoughts on its future. IEEE Trans Eng Manag 41(4):342–346
8. Nambisan S, Wilemon D (2003) A global study of graduate management of technology programs. Technovation 23:949–962
9. Pilkington A (2006) Conceptualizing the management of technology. In: Proceedings of the 2nd European conference on management of technology: technology and global integration, Birmingham, UK
10. Pilkington A (2008) Engineering management or management of technology? A bibliometric study of IEEE TEM. Int J Manag Sci Eng Manag 3(1):63–70
11. McMillan GS (2008) Mapping the invisible colleges of R&D Management. R&D Manag 38(1):69–83
12. Biemans W, Griffin A, Moenaert R (2007) Twenty years of the Journal of Product Innovation Management: history, participants, and knowledge stock and flows. J Prod Innov Manag 24:193–213
13. Steven W, Xielin L, Wei X (2001) A survey of Chinese literature on the management of technology and innovation, 1987–1997. Int J Technol Manag 21(1, 2):130–145
14. Seol S, Park J (2008) Knowledge sources of innovation studies in Korea: a citation analysis. Scientometrics 75(1):3–20
15. Pritchard A (1969) Statistical bibliography or bibliometrics. J Documentation 25:348–349
16. Usdiken B, Pasadeos Y (1995) Organizational analysis in North America and Europe: a comparison of co-citation networks. Organizational Stud 16(3):503–526
17. Culnan M (1986) The intellectual development of management information systems, 1972–1982: a co-citation analysis. Manag Sci 32(2):156–172
18. Pilkington A, Liston-Heyes C (1999) Is production and operations management a discipline? A citation/co-citation study. Int J Oper Prod Manag 19(1):7–20
19. Ramos-Rodriguez AR, Ruiz-Navarro J (2004) Changes in the intellectual structure of strategic management research: a bibliometric study of the Strategic Management Journal, 1980–2000. Strateg Manag J 25:981–1004
20. Nerur SP, Rasheed AA, Natarajan V (2008) The intellectual structure of the strategic management field: an author citation analysis. Strateg Manag J 29:319–336
21. Martinsons MG, Everett JE, Chan K (2001) Mapping the scholarly development of strategic management. J Inf Sci 27(2):101–110
22. Acedo FJ, Casillas JC (2005) Current paradigms in the international management field: an author co-citation analysis. Int Bus Rev 14:619–639
23. Gu Y (2004) Global knowledge management research: a bibliometric analysis. Scientometrics 61(2):171–190
24. Cornelius B, Persson O (2006) Who's who in venture capital research. Technovation 26:142–150
25. Adler PS (1989) Technology strategy: a guide to the literatures. In: Rosenbloom R, Burgelman R (eds) Research in technological innovation, management and policy, vol 4. JAI Press, Greenwich
26. Drejer A (1997) The discipline of management of technology, based on considerations related to technology. Technovation 17(5):253–265

27. Allen JT, Varghese G (1989) Changes in the field of R&D Management over the past 20 years. R&D Manag 19(2):103–113
28. Beard JW (2002) Management of technology: a three-dimensional framework with propositions for future research. Knowl Technol Policy 15(3):45–58
29. Liao S (2005) Technology management methodologies and applications: a literature review from 1995 to 2003. Technovation 25:381–393
30. Ball DF, Rigby J (2005) Disseminating research in management of technology: journals and authors. R&D Manag 36(2):205–216
31. Linstone HA (1999) TFSC: 1969–1999. Technol Forecast Soc Chang 62:1–8
32. Callon M, Coenen R, Cohen WM, Freeman C, Kodama F, Meyer-Krahmer F, Pavitt K, Pisano G (1999) Retrospective evaluation (1971–1999). Res Policy 28:911–919
33. Allen T, Sosa L (2004) Fifty years of engineering management through the lens of the IEEE Transactions. IEEE Trans Eng Manag 51(4):391–395
34. Merino MTG, Carmo MLP, Alvarez MVS (2006) 25 years of Technovation: characterisation and evolution of the journal. Technovation 26:1303–1316
35. Cheng CH, Kumar A, Motwani JG, Reisman A, Madan MS (1999) A citation analysis of the technology innovation management journals. IEEE Trans Eng Manag 46(1):4–13
36. Liker J (1995) Results of survey of management journals for TIM research. TIM Newsletter 7(2):5–8
37. Franske N, Schreier M (2005) A meta-ranking of TIM and entrepreneurship journals: aggregation and calibration to a common metric. Working paper Vienna University of economics, Vienna
38. OECD (2008) Country classification. Available at http://www.oecd.org/dataoecd/53/15/42203230.pdf
39. Schildt H (2005) Sitkis—a bibliometric software tool. Espoo, Finland
40. Borgatti SP, Everett MG, Freeman LC (2002) Ucinet for windows: software for social network analysis. Analytic Technologies, Harvard
41. Garfield E (1979) Is citation analysis is a legitimate evaluation tool? Scientometrics 1(4): 359–375
42. Small HG (1978) Cited documents as concept symbols. Soc Stud Sci 8(3):327–340
43. Small HG (1973) Co-citation in the scientific literature: a new measure of the relationship between two documents. J Am Soc Inf Sci 24(4):265–269
44. White HD, Griffith BC (1981) Author co-citation: a literature measure of intellectual structure. J Am Soc Inf Sci 32(3):163–171
45. Small HG (1977) A co-citation model of a scientific specialty: a longitudinal study of collagen research. Soc Stud Sci 7:139–166
46. de Solla Price DJ (1963) Little science big science. Columbia University Press, New York
47. Crane D (1972) Invisible colleges: diffusion of knowledge in scientific communication. University of Chicago Press, Chicago
48. Knoke D, Yang S (2008) Social network analysis. Sage Publications, Los Angeles
49. Cohen W, Levinthal D (1990) Absorptive capacity: a new perspective on learning and innovation. Adm Sci Q 35:128–152
50. Nelson RR, Winter SG (1982) An evolutionary theory of economic change. Belknap Press, Cambridge
51. Barney J (1991) Firm resources and sustained competitive advantage. J Manag 17(1):99–120
52. Prahalad CK, Hamel G (1990) The core competence of the corporation. Harvard Bus Rev 68:79–91
53. Kim L (1997) Imitation to innovation: the dynamics of Korea's technological learning. Harvard Business School Press, Boston
54. Porter ME (1980) Competitive strategy: techniques for analyzing industries and competitors. Free Press, New York
55. Porter ME (1985) Competitive advantage: creating and sustaining superior performance. Free Press, New York
56. Porter ME (1990) The competitive advantage of nations. Free Press, New York
57. Kogut B, Zanger U (1992) Knowledge of the firm combinative capabilities, and the replication of technology. Organ Sci 3(3):383–397

58. Leonard-Barton D (1992) Core capabilities and core rigidities: a paradox in managing new product development. Strateg Manag J 13:111–125
59. Leonard-Barton D (1995) Wellsprings of knowledge: building and sustaining the sources of innovation. Harvard Business School Press, Boston
60. Huber GP (1991) Organizational learning: the contributing processes and the literatures. Organ Sci 2(1):88–115
61. Nonaka I (1995) Dynamic theory of organizational knowledge creation. Organ Sci 5(1):14–37
62. Nonaka I, Takeuchi H (1995) The knowledge-creating company: how Japanese companies create the dynamics of innovation. Oxford University Press, New York
63. Grant RM (1996) Toward a knowledge based theory of the firm. Strateg Manag J 17:109–122
64. Nelson RR (1993) National innovation systems: a comparative analysis. Oxford University Press, New York
65. Castellacci F (2006) A critical realist interpretation of evolutionary growth theorizing. Camb J Econ 30(6):861–880
66. Mulder P, De Groot HLF, Hofkes MW (2001) Economic growth and technological change: a comparison of insights from a neo-classical and an evolutionary perspective. Technol Forecast Soc Chang 68:151–171
67. Lundvall BA (1997) National systems of innovation: towards a theory of innovation and interactive learning. Pinter Publishers, London
68. Edquist C (1997) Systems of innovation: technologies, institutions and organisations. Pinter Publishers, London
69. Dosi G (1982) Technological paradigms and technological trajectories. Res Policy 11:147–162
70. Freeman C (1987) Technology policy and economic performance: lessons from Japan. Pinter Publishers, London
71. Bell M, Pavitt K (1993) Technological accumulation and industrial growth: contrasts between developed and developing countries. Ind Corporate Change 2:157–210
72. Bell M (1984) Learning and the accumulation of industrial technological capacity in developing countries. In: King K, Fransman M (eds) Technological capability in the third world. Macmillan, London
73. Bell M, Pavitt K (1995) The development of technological capabilities. In Haque IU (ed) Technology and international competitiveness. The World Bank, Washington
74. Lall S (1992) Technological capabilities and industrialization. World Dev 20(2):165–186
75. Amsden AH (1989) Asia's next giant: South Korea and late industrialization. Oxford University Pres, New York
76. Lee JJ (1988) Technology development processes a model for a developing country with a global perspective. R&D Manag 18(3):235–250
77. Zahra SA, George G (2002) Absorptive capacity: a review, reconceptualization, and extension. Acad Manag Rev 27(2):185–203
78. Kim L (1998) Crisis construction and organizational learning: capability building in catching-up at Hyundai motor. Organ Sci 9:506–521
79. Kim L (1999) Building technological capability for industrialization: analytical frameworks and Korea's experience. Ind Corporate Change 8(1):111–136
80. Baruch Y (2001) Global or North American? A geographically based comparative analysis of publications in top management journals. Int J Cross Cult Manag 1(1):109–126
81. Freeman C (1974) The economics of industrial innovation, 1st edn. Penguin, Harmondsworth
82. Rogers EM (1962) Diffusion of innovations. 1st (edn) Free Press, New York
83. Utterback JM (1994) Mastering the dynamics of innovation: how companies can seize opportunities in the face of technical change. Harvard Business School Press, Boston
84. Utterback JM, Abernathy WJ (1975) A dynamic model of product and process innovation. Omega 3(6):639–656
85. Eisenhardt KM (1989) Building theories from case study research. Acad Manag Rev 14(4):532–550
86. Teece DJ (1986) Profiting from technological innovation: implications for integration, collaboration, licensing and public policy. Res Policy 15(6):285–305

# Chapter 3
# Ontology- and Function-Based Technology Model for Decision Making in New Product Development

**Stefanie Bunzel, Joachim Warschat, Dieter Spath and Antonino Ardilio**

**Abstract** One of the key factors for a successful product development is the early integration of technologies in the innovation process. Especially emerging technologies, as e.g. nanotechnologies, pose a wide range of new functionalities and therefore high potentials for future products. To identify, which technological knowledge meet the product requirements in the early stages of the product development, it is necessary to create a connection between those two domains. This connection is realized by a function-based approach and provides product developer a systematic access to fitting technological solutions. By using an ontology-based model for the technology domain, it is possible to interrelate all the relevant aspects (functionalities, material, structures, sizes, etc. as well as corresponding experts and research institutions) and therefore provide a comprehensive and logically connected data pool for decision making. The approach presented in this chapter is especially tailored for the particular needs of new product developer. The underlying IT support allows fats and extensive search as well as the extraction of the adequate solutions within this data pool for product development. The implementation and combination with specific product requirements as well as the decision making process will be shown using the example of nanotechnology.

S. Bunzel (✉) · J. Warschat · D. Spath · A. Ardilio
Fraunhofer Institut für Arbeitswirtschaft und Organisation IAO (Fraunhofer Institute for Industrial Engineering), Stuttgart, Germany
e-mail: stefanie.bunzel@iao.fraunhofer.de

D. Cetindamar et al. (eds.), *Strategic Planning Decisions in the High Tech Industry*, DOI: 10.1007/978-1-4471-4887-6_3,

## 3.1 Nanotechnology Applications for New Product Ideas

### *3.1.1 Characteristics of Nanotechnology*

Structures and materials that are smaller than 100 nm and that show new properties and effects that cannot be seen in the macro world are denotative for the nanotechnology [1–3]. These effects can be categorized in chemical, mechanical, electrical, biological, thermal, geometrical, optical and magnetic effects. Along with the great potential that nanomaterials and -structures poses for enhanced or new product properties, there are mainly two aspects that characterize nanotechnology in a particular way:

- The development and the range of applications of nanotechnology are subject to great dynamics, diversity and complexity [4, 5] where upon there are gaps between products and technology development, a so-called "valley of death" [6, 7]. Reasons therefore are among other things a missing explicit vision from possible users of what nanotechnology might accomplish if used in their products [8–11].
- The new properties of nanomaterials necessitate a systematic approach to be able to better control the complexity and at the same time the universality on the material level and to allow for a methodical evaluation of the relevant properties and materials [12–16]. Similar product properties can be created by different materials and coatings. Or a single material shows several different properties like for example the CNTs (see following Sect. 3.1.2).

### *3.1.2 CNTs as an Example of Nanomaterials*

The potential of nanotechnology can be demonstrated very well with the material class "carbon nanotubes" (CNT). CNTs are single-walled (SWCNT) or multi-walled (MWCNT) tubes composed of carbon with a diameter of several nanometers and an aspect ratio of 100 to 10.000. The chirality is responsible for the conductive and semi conductive properties of the CNTs. The manufacturing technology for the CNTs is the arc discharges between carbon electrodes—the catalytic decomposition of carbonaceous gas in an oven. This technology is called the chemical vapour deposition or laser vaporization of graphite. Within this material class there is a multiplicity of material- structural designs, that enable different combinations of working structures (nanotechnological structure (nanomaterials or nano-structure), which carries the specific function(s) as an applicable system) and function (specific function, based on nano technological effects of nanomaterials). For the future the CNTs offer a very high economic potential due to their exceptional and their unique structural, mechanical, electrical, thermal, optical, chemical and sensory- actuatory properties.

## 3.2 Ontologies for the Illustration of a Knowledge Domain

For the formalization and the illustration of the technological functions and the product functions semantic networks and ontologies are being used. Those technological functions are dynamically usable, in comparison with established approaches like the "technology calendar" or "construction catalogues".

In the last two decades the term ontology has gained great significance in particular within the framework of knowledge management and informatics. However, the term is not based on a philosophical understanding, but rather on the works and findings in the field of artificial intelligence. In the following some of the predominantly definitions in the relevant literature shall be analyzed to identify the basic aspects that determine this "modern" ontology term. A very early definition of the term ontology within the meaning of informatics was published in 1991 by Neche [17] in the AI Magazine: "An ontology defines the basic terms and relations comprising the vocabulary of a topic area as well as the rules for combining terms and relations to define ex-tensions to the vocabulary".

This definition already refers to some of the most essential elements of ontology. According to this the vocabulary is being defined to a certain knowledge domain. In addition it becomes apparent that this vocabulary is being composed of the essential terms of the domain as well as the relation between those terms. Also this definition describes the rules on how to combine the terms with each other to be able to expand the vocabulary. However, this definition does not pose any immediate conclusions on the form in which this vocabulary should be illustrated. One of the best known and most cited definitions on ontologies was verbalized by Gruber in 1993 [18]: "An ontology is an explicit specification of a conceptualization" and "a conceptualization is an abstract, simplified view of the world that we wish to represent for some purpose".

According to Gruber ontologies are abstract models, which contain all the relevant terms of a domain that are needed to illustrate a specific purpose. Moreover all those abstract models have to be described in an explicit specification, which means that the denotation of all the terms within the models have to be clear and well-defined in a structural way. With this Gruber not only emphasizes on the conceptual aspect of ontologies, but also addresses the pragmatic component. In the context of computer sciences no statements are being made with ontologies regarding the "being" as such or the aim to build up the reality. Rather ontologies can be understood as "artifacts", consisting of a well-defined vocabulary for the description of a reality that is being aligned on a specific purpose. Gruber's definition is not only one of the most cited but also one of the most discussed definitions in the according literature.

Many authors generally agree with Gruber, but often they consider his interpretations as too universally valid. A great many of further terminology therefore developed with the aim to expand the construction of Gruber. To this group the following two definitions from Borst and Karp can be added as well: "Ontologies are defined as a formal specification of a shared conceptualization" [19]. "An ontology is a specification of a conceptualization that is designed for reuse across multiple applications and implementations. Put another way, a specification of a conceptualization is a written, formal description of a set of concepts and relationships in a domain of interest" [20].

Both of the authors expand the definition of Gruber by two essential aspects. On the one hand they emphasize on the formal linguistic aspect of ontologies. Consequently, ontologies have to be written in a machine-readable language. On the other hand they add the requirement that there has to be a consensus within the group about the formalized knowledge ("shared"). Without the general acceptance on the part of the users, an ontology is not applicable or reusable in another context ("designed for reuse"). Ontologies are established in the knowledge engineering for the formalization of representations and for the exchange of knowledge.

The first successful application areas of ontologies are e.g. in the medical industry. The existing knowledge concerning the medical research purpose is being illustrated formalized in ontologies. Another example is the field of mechanical engineering, where an ontology-based semantic network was successfully established as a homogeneous data basis for rapid prototyping [21]. Further ontology based applications can be found in academic knowledge portals, in the data management and system integration, in the electronic commerce, for the planning of production systems or for semantic web services in general [22–26].

In addition to the illustration of technological knowledge ontologies also offer advantages by simplifying the related identification of experts and to connect them with the technological knowledge. By using adequately modeled ontologies new knowledge can be generated by inference.

To be more specific: It is possible to search for technologies and applications in the ontology that comply with very specific criteria. That way the ontology model of an expert can be combined with the model of one or more knowledge fields and offers the possibility to realize the search for experts. This approach is not the combination of technology key words but it is rather problem-orientated. In the first step this means, that the relevant technology is being identified in the ontology by certain criteria and in a secondary search the associated experts are being identified immediately (see Fig. 3.1). With this approach the search for experts becomes much more problem-oriented and therefore more efficient.

## 3.3 Field of Research

In the domain of nanotechnology a conceptual approach is needed, that allows for a way to bring together all the dynamically appearing research findings and information in the respective fields in an interrelated and logical context [27, 28]. Especially the increasingly complex crosslinking of the involved disciplines chemistry, physics and biotechnology and the inadequate standardized terminologies and definitions exceed the degree of manageable complexity by using

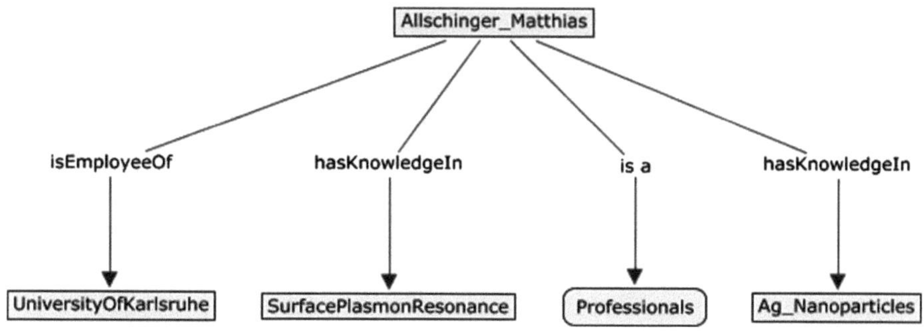


**Fig. 3.1** Illustrates the connection between experts and the technological knowledge domain

conventional methods of formalization like e.g. tabular data basis, topic maps or conventional knowledge management models.

Those approaches possess the following main deficits: low flexibility pertaining to the maintenance and actuality of the knowledge, the missing crosslinking of the knowledge and the partly inadequate machine readability. In the case of nanotechnology it is possible to illustrate e.g. information in a structural way in a data basis and to allocate definite correlations, but it is not possible to draw logical conclusions out of the entity of the data. Therefore an approach for the formalization is required that can edit and process a knowledge domain with a formal logic as well as machine-readable to be able to refer heterogeneous data on different levels and detail degrees to each other.

Examples for those references are the description of properties and new functionalities, performance features, the illustration of specific working principles or the allocation of competencies of research institutes and technical experts. There are high requirements for the formalization approach regarding the accuracy of the search results. Simultaneously new possible solutions for the existing product requirements shall be identified on the basis of the technology and through the logical crosslinking of the information.

## 3.4 Concept for an Ontology-Based Approach to the Field of Nanotechnology

The below described concept comprises the use and integration of new approaches from the fields of digital engineering, construction, identification of product ideas and semantic concepts.

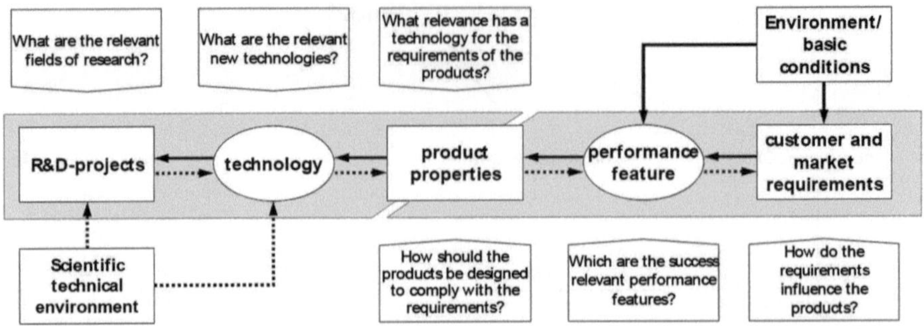


**Fig. 3.2** Value chain regarding the technology analysis

### *3.4.1 Approach for the Analysis of Emergent Technologies for New Product Ideas*

In the context of the planning and composition of new product ideas the developer has to analyze and evaluate the relevance of different technologies according to the requirements. The fundamental question at the beginning is whether a technology can provide the requested function or not. To answer this question the technology relevance analysis is a suitable method a detailed description of the subject can be found at [29, 30].

The approach of the technology relevance analysis consists of an interplay between technological possibilities (technology push) and the requirements of the market (market pull) (see Fig. 3.2). It is the aim to harmonize the new technical possibilities on the one hand with the customer and market requirements on the other hand.

### *3.4.2 Basic Principle of the Concept*

To improve the interoperability, to so-called faculty of speech between the two knowledge domains "nanotechnology" and "product development", in this concept a standardized formalization approach is being chosen to be able to describe the two knowledge domains on the basis of functions. The technology domain is being described by solution-neutral working principles that can be derived from the possible properties and functionalities as well as from the working structure of the technology. This approach resembles the conventional approach long used in the construction.

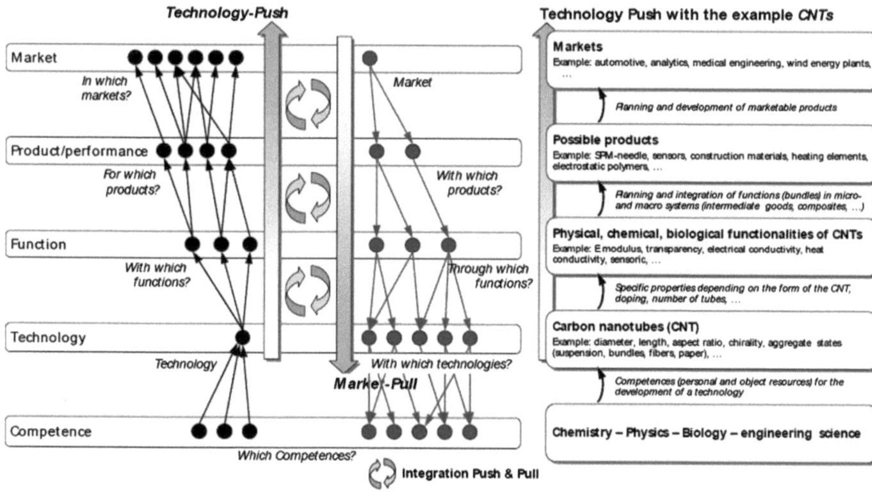


**Fig. 3.3** Level model for the combination of nanotechnology and applications through aim-means-combinations

## 3.4.3 *Functions for the Enabling of Aim-Means Combinations*

The classification of the functional types in user and product related functions ("purpose function" and "system function") allows for the constitution of an aim-means combination:

- A purpose function is a user-related function that describes the expected or adduced effect of a product with the aim to fulfill a user requirement.
- A system function is a product-related function that is used for the fulfillment of user-related functions through the effect of a product element or between the component parts.

Correspondingly a purpose function can be realized by different system functions (which mean solution principles). Those "solutions in principle" demonstrate the possibility of realizing a user-related function in a simplified way through the working principle and the associated structure. The functional dimension with the purpose function and the system function serves thus as a connector between the levels product and technology [29, 31]. Figure 3.3 shows the complexity of the sought-after aim-means-connections: Nanomaterials provide different (system) functions, which can be applied in quite different products as purpose function. And vice versa a product as a matter of principle can use quite different functionality of the nanotechnology, which means that the product contains Nanomaterials. Using the functional dimension the relevant solutions paths can be recognized und describes. That way it serves as a "hinge" between the product and the nanotechnology [32].

**Table 3.1** Overview on the functions of nanotechnology as well as CNT specific properties

| Functionalities of nanotechnology | Generally improved and new usable properties due to nanotechnology | CNT-specific properties |
|---|---|---|
| Chemical | Hardness, breaking strength | Density: 1,3 (SWCNT)—2,6 (MWCNT) g/cm$^3$; E modulus: 563 (SWCNT-bundle)—1,200 (MWCNT) GPa |
| Biological | Chemical reaction/reactivity/enlarged surface energy, sintering activity, bonding selectivity, specific selectivity and reaction as well as chemical stability | Storage of hydrogen catalyzer carrier molecular filtering |
| Optical | Interaction of nanoparticles with biological systems, membrane mobility of nanoparticles, biological addressing and masking of nanoparticles, selforganization processes | Bio-imprinting: Fabrication of structures with recognition functions for bio molecules |
| Magnetic | Conductivity, isolation, thermochrome phases | Distinctive heat conductivity |
| Electrical | Additional electronic phases (quantum dot or quantum well phases), electronic surface phases influence the electronic properties, quantization of the electrical charge | Metallic conductive or semi conductive (dependent on structure and diameter) |
| Thermal | (Super) paramagnetism, ferromagnetic properties, magneto resistive effects (XMR) | Very good magnetic and electro magnetic properties |
| Mechanical | Transparency, reflectance, scattering effect, light absorption/light emission of a certain wavelength range | Very accurate measurement through the use of scanning probe needles |

#### 3.4.3.1 Description of the Technological Possibilities of Nanotechnology Through Working Principles

The effect of nanotechnology as an aim-means combination is being composed of materials and its functions. According to this a working principle can be described as an interaction (= function) and a target area (= material):

The interaction indicates the aim and therefore the effect that can be realized in principle. The target area describes the means and therefore the location where the interaction takes place. Initially this can be a (contact) surface or a space.

The effect of nanotechnology can be described by seven different nano-functions as can be seen in Table 3.1.

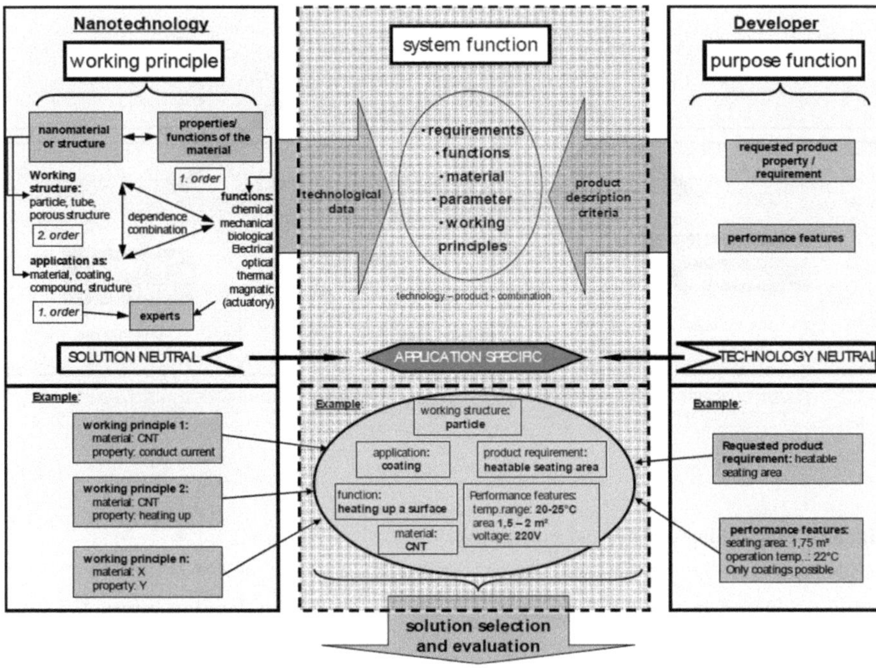


**Fig. 3.4** Coupling of the working principle and the user-related function

#### 3.4.3.2 Semantic Illustration of Nano and New Product Knowledge Domains with Ontologies

For the formalization and illustration of the functions the possibilities of semantic networks and ontologies are being used. With the help of semantic modeling the respective working principles of a nanotechnology can be interrelated in a logical context so that the basis for a translation into the product level is being set. Figure 3.4 illustrates the theoretic concept of the ontology-based model.

The main aim of the development of the semantic model is the integration of new technologies into the product development at an early stage through an increased interoperability between the associated knowledge domains of the technology management. With this approach, product developers are able to get an operationalized access to the applicability of emerging technologies in the context of the specific product conception.

For the semantic modeling of this theoretic concept (see Fig. 3.4), it is necessary to divide the model into three main parts (see Fig. 3.5). On the left side an ontology for the entire knowledge domain of nanotechnology has to be built up using a software tool, e.g. the tool "Protégé", an open source ontology editor and knowledge-base framework developed by the Stanford University School of Medicine. As description language the web ontology language OWL was used. For the pre-population of the ontology as well as for the intuitive navigation and the

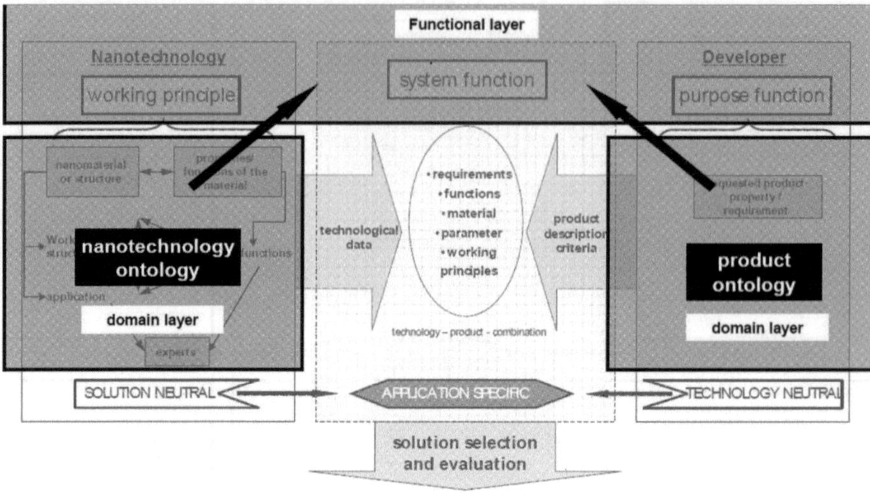


**Fig. 3.5** Establishing domain and functional layers

semantic search in the solution areas it is also necessary to develop interactive user interfaces.

On the left side a semantic model of product-related requirements and performance features has to be established (see Fig. 3.5). The basis for the semantic model are the existing functional catalogs, e.g. from [33–36].

Those two ontologies are within the domain layer. To bring both domains together, another layer has to be established—the so-called functional layer (middle column in Fig. 3.5). This is the connection point of the two domains "nanotechnology" (left column in Fig. 3.5) and "product" (right column in Fig. 3.5) and offers on basis of the system functions solutions for new product ideas. The main advantage of this framework is the possibility of replacing technology domains with others without having to rebuild the entire system.

Figures 3.6 and 3.7 show a concrete semantic illustration of an ontology using the example of nanotechnology—in a schematic form and as an extract from the software tool Protégé. Besides the illustration of material properties it is also possible to integrate selected experts with their competences into the ontology to provide a contact person for deeper matter within the selected solutions.

### *3.4.4 Concept Implementation in an IT Tool*

For the implementation of the concept in practice, an ontology-based approach for a combined functional view of application and nanotechnology was implemented in a software tool. The aim is the systematic and semi-automatic support of the concept for the product development. Therefore, it is necessary to bring together

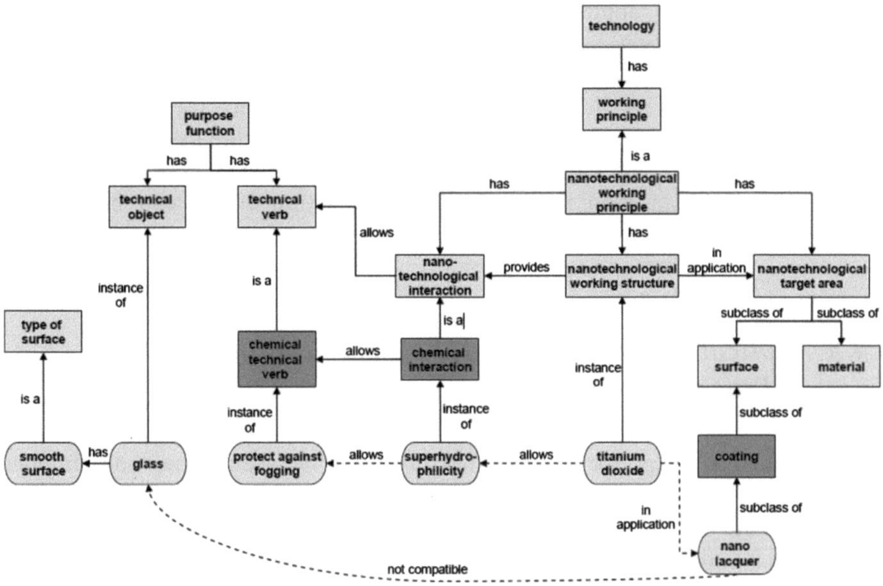


**Fig. 3.6** Extract of a schematic ontology using the example of nanotechnology

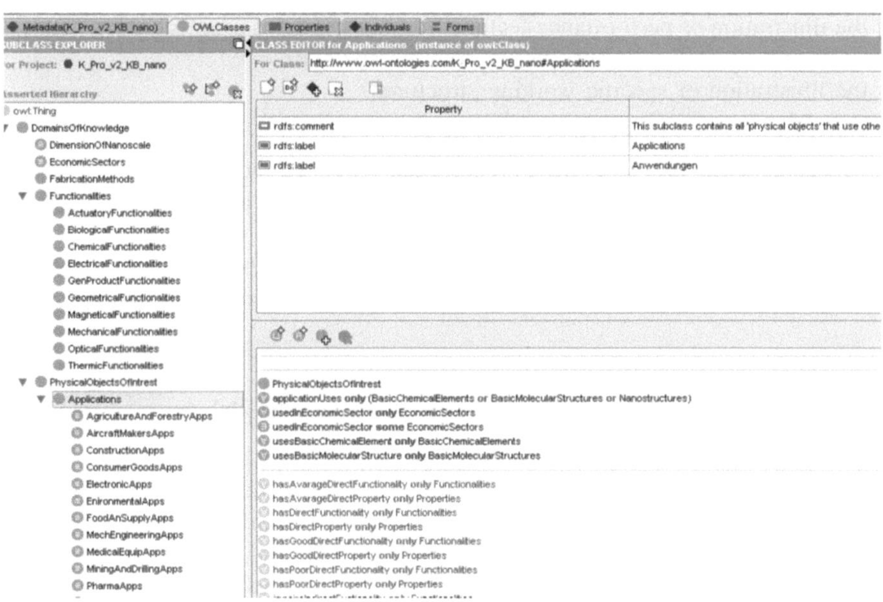


**Fig. 3.7** Extract of the nanotechnology ontology in Protégé

adequate description approaches for technology working principles and the favored purpose functions as well as design corresponding connection rules for the illustration of the system functions. Complex and logical relations can then be made analyzable and usable through reasoning and inference formation.

In excess of the illustration of already known relations, the system also should be able to suggest unknown or not explicitly formulated relations through the logical model. In addition the information extraction could also be possible through the introduction of the approach of "user back coupling".

### *3.4.5 Application Potential and the Benefit of the Concept in Practice*

The presented concept for a combined language between nano- and product-domain with the aim to better connect research and application of the nanotechnology uses the functional dimension und distinguishes between purpose functions and system functions that are based on the working principles of nanotechnology. Therefore, the abstract functional level serves as the hinge between product and nanotechnology. The range of applications of the presented concept in practice is very comprehensive and offers product developer support for the following fields:

- the description of properties and new functionalities of the nanotechnology,
- the illustration of performance features for nanoparticles as a technology for the application (solution unspecific vs. solution specific),
- the illustration of specific working structures,
- the allocation of competences of research institutes and technical experts.

Hence concrete problems from practice can be solved; however, the application of this concept goes even further. This concept could also support creativity processes by illustrating the "spacial closeness" of terms in regard to the content in interactive graphics. By the intuitive browsing through the semantic interrelated concept field, new or alternative possible solutions could unfold to the user.

## 3.5 Information Visualization for Decision-Making

To visualize the contents of an ontology for specific tasks appropriate visualization techniques are needed. The users of an ontology-based information system need visualization tools that not only display simple information, but visualization tools through which interactive discussion with the stored knowledge within an ontology becomes possible. In the following some visualization tools are being presented, allowing the analysis of the knowledge base with different objectives. The overall goal will be to identify suitable visualization methods for a specific

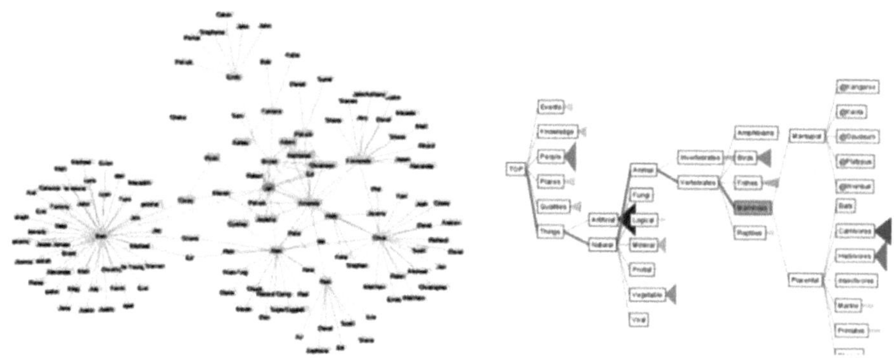

**Fig. 3.8** Different facets of Node-Link diagrams: force-directed layout (*left*), space tree (*right*)

OWL-ontology to be used in decision making for product development (evaluation of found solutions from emergent technologies in the context of requirements management).

## *3.5.1 Node-Link Diagrams*

For the illustration of network structures or hierarchic structures, node-link diagrams often are used (Fig. 3.8). Various differences appear in respect of the layout for the illustration of the knots of a graph and its configuration. Geometric forms like (e.g. circles or rectangles) are deployed to represent an element (class or individual).

## *3.5.2 Focus and Context*

These techniques pursuits the goal to combine the context views of an element with the corresponding detail view. The user can concentrate on individual objects (e.g. classes or individuals) without losing sight of the context (e.g. higher ranked classes). Two of those visualization methods are the hyperbolic tree (HT) from Xerox PARC and the MoireGraph from Tricot and Roche [34] (both see Fig. 3.9).

Those techniques project the hierarchy onto the surface of a ball. Thereby a hyperbolic geometry is being used instead of an Euclidean. In spite of a high number of knots, the context stays within the field of vision even when a new knot is being pushed into the center of the illustration by an user interaction [37].

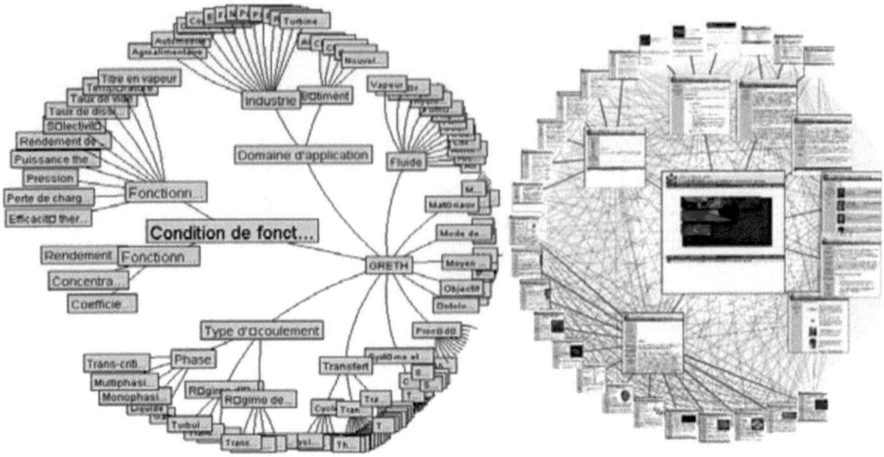


**Fig. 3.9** Hyperbolic Tree (*left*) and MoireGraph (*right*)

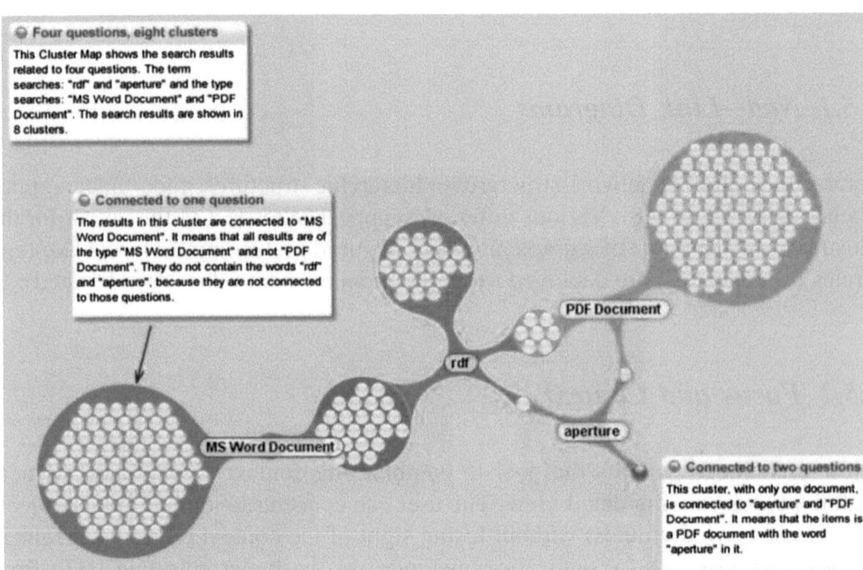


**Fig. 3.10** Example of a cluster map (http://www.aduna-software.com)

### *3.5.3 Cluster Maps*

This method (introduced by the company Aduna) bases on the "spring embedding algorithm" and was developed to visualize the hierarchical relations of the classes and individuals of semantic networks. Elements can be positioned next to each other due to their semantic closeness (Fig. 3.10). The more individuals a class has, the closer are those clusters in the illustration. Fluit [38] demonstrate that cluster maps exhibit a high expressivity due to the fast comprehending relations within the taxonomy [38]. The instrument "Aduna Autofocus" uses listings and input fields as support for the user.

## 3.6 Conclusions and Outlook

By means of the presented concept for a combined functional dimension of applications and nanotechnology with the aim to have a support for new product ideas, technical knowledge is being interrelated with product requirements that have been identified in the early stages of development. Thus, it is possible to point out and qualitatively evaluate the potential of found solution, based on emerging technologies for a product developer.

The concept bases on a standardized formalization approach in terms of a function-based description of the knowledge domains on the basis of technological data and product requirements. A functional dimension is being used comprising purpose functions (requirements regarding the product) and system functions (functions provided by nanotechnology in the form of working principles). For the formalization and the illustration of the functions semantic networks and ontologies are being used, since they are more dynamically usable than established approaches like the technology calendar or construction catalogs.

With this presented approach relevant knowledge domains can be processed with a formal logic and machine-readable manner. Thus, product developers are able to relate heterogeneous data on different levels and detail layers in order to support decision making ("What technology solutions are existing and which one fits best to the requirements").

Besides the implementation of the concept into a software tool using information technologies and its validation in practice, there are further challenges. It is necessary to develop sustainable concepts for the still open question regarding the continuous updating of the technological knowledge data, for example by integrating technology experts. Furthermore there is research needed concerning automated operations for text analysis and web crawler [39] to be able to evaluate technology information and to transform them—at least semi-automatically—into ontologies in order to include new data into the existing knowledge basis.

## References

1. Bundesministerium für Bildung und Forschung (Hrsg.) (2006) Nano-Initiative—Aktionsplan 2010. Bonn, Berlin http://www.bmbf.de/pub/nano_initiative_aktionsplan_2010.pdf
2. Heubach D, Beucker S, Lang-Koetz C (2009) Einsatz von Nanotechnologie in der hessischen Umwelttechnologie—Innovationspotenziale für Unternehmen. (HessenAgentur, Hrsg.) Band 1 der Schriftenreihe der Aktionslinie Hessen-Nanotech des Hessischen Ministeriums für Wirtschaft, Verkehr und Landesentwicklung
3. Pastewski N, Lang-Koetz C, Heubach D (2009) Materialeffizienz durch den Einsatz von Nanotechnologien und neuen Materialien. (HessenAgentur, Hrsg.) Band 14 der Schriftenreihe der Aktionslinie Hessen-Nanotech des Hessischen Ministeriums für Wirtschaft, Verkehr und Landesentwicklung
4. Luther (2004) Nanotechnologie als wirtschaftlicher Wachstumsmarkt: innovations- und Technikanalyse (V. Technologiezentrum, Hrsg.) Zukünftige Technologien
5. Shea CM (2005) Future management research in directions in nanotechnology: a case study. In: journal of engineering and technology management
6. Fuchs H, Grandke T, Grünhagen H, Gutsch A, Heubah D, Iden R (2007) Große potenziale der nanowelt ergreifen—rücken schlagen, chancen umsetzen, verantwortlich handeln, impulskreis nanowelten in der initiative partner für innovation- Bilanz des arbeitsjahres. Stuttgart
7. Markham SK (2002) Moving technologies from lab to market. Res Tech Manage 45(6):30–41
8. Brune H, Ernst H, Grunwald A, Grünwald W, Hofmann H, Krug H, Janich P, Mayor M, Rathgeber W, Schmid G, Simon U, Vogel V, Wyrwa D, Mader K (2006) Nanotechnology—assessment and perspectives. Berlin, Heidelberg
9. Agentur H (2005) Nanotechnologie in Hessen—ein bestandsaufnahme auf basis von unternehmensbefragungen, Wiesbaden
10. Kingon A (2004) Corporate responses to nanoscience and nanotechnology. Research technology management
11. Zweck A (2005) Technologiemanagement—technologiefrüherkennung und technikbewertung. In: Schäppi B, Andreasen M, Kirchgeorg M, Radermacher FJ (ed) Handbuch Produktentwicklung. München, Wien
12. Decker M, Fiedeler U, Fleischer T (2004) Ich sehe was, was Du nicht siehst zur Definition von Nanotechnologie Technikfolgenabschätzung—Theorie und Praxis, S. 10–16
13. Fleischer T, Grunwald A (2005) Innovationen in neuen Technologiefeldern—Die Nanotechnologien. WSI Mitteilungen 58
14. Paschen C, Coenen T, Fleischer R, Grünwald D, Oertel C (2004) Nanotechnologie: forschung, entwicklung, Anwendung, Berlin, Heidelberg
15. Rathgeber W, Simon U, Wyrwa D (2003) Small dimensions and material properties—a definition of nanotechnology
16. Spath D, Warschat J, Heubach D (2007) An approach for a relevance analysis of nanotechnology. Design for Society—innovation, sustainability and knowledge
17. Neches R, Fikes R, Finin T, Gruber T, Patil R, Senator T, Swartout W (1991) Enabling technology for knowledge sharing, Al magazine
18. Gruber T (1993) A translation approach to portable ontologies. Knowledge Acquisition
19. Borst W (1997) Construction of engineering ontologies. University of Tweenty, Enschede, NL: centre for telematica and information technology
20. Karp PD (2000) An ontology for biological function based on molecular interactions, In: bioinformatics ontology
21. Diederich M, Warschat J. (2007) Wissensrepräsentation und Kommunikation. Entwicklung und Erprobung innovativer Produkte—Rapid Prototyping
22. Fischer T, Murphy M, Tippmann J, Ayroumlou M (2005) Semantic web services enabling collaborative engineering, 11th international conference on human-computer interaction. Las Vegas, USA

23. Gao and Roller (1998) Semantic based product model: principles and representations. In: Roller D (ed) Proceedings of 31st ISATA, volume automotive mechatronics design and engineering, Croydon, England (Hrsg)
24. Schlegel T, Thiel S, Lotterbach S (2007) Produktion 2.0—semantic web services in der produktion. Zeitschrift für wirtschaftlichen Fabrikbetrieb
25. Spath D (2005) Production engineering: research and development in Germany, Annals of the german academic society for production engineering, Jg. 12
26. Sure Y, Staab S, Studer R (2001) On-to-knowledge methodology. Von ontoknowledge.org: employed and evaluated version, On-to-knowledge EU IST-1999-10132 Project Deliverable D16 (WP5), http://www.ontoknowledge.org/downl/del16.pdf. Abgerufen
27. Gerybadze A (2004) Technologie- und Innovationsmanagement: strategie, organisation und implementierung, München
28. Spielberg D (2002) Methodik zur Konzeptfindung basierend auf technischen Kompetenzen, RWTH Aachen
29. Heubach D (2008) Eine funktionsbasierte Analyse der Technologierelevanz von Nanotechnologie in der Produktplanung, IPA-IAO Forschung und Praxis 478. Heimsheim
30. Ardilio A, Spath D, Warschat J (2009) The potential of emerging technologies: strategy-planning for technology-providers throughout an application-radar, management of engineering and technology, PICMET 2009. Portland international conference on management of technology, 462–477
31. Bullinger H (2008) Der Technologieentwicklungsprozess. In: Bullinger HJ, Fokus Technologie. Chancen erkennen, Leistungen entwickeln (Hrsg)
32. Peiffer S (1992) Technologie-Frühaufklärung, Hamburg
33. Birkhofer H (1980) Analyse und Synthese der Funktionen technischer Produkte. Fortschrittsberichte der VDI Zeitschriften Nr. 70
34. 1325-1:1996-11, DIN DE (1996) Value Management, Wertanalyse, Funktionenanalyse, Wörterbuch—Teil 1: Wertanalyse und Funktionenanalyse, Deutsche Fassung EN 1325-1:1996
35. VDI-2221: VDI-Richtlinie 2221(1993) Methodik zum Entwickeln und Konstruieren technischer Systeme und Produkte, VDI-Gesellschaft Produkt- und Prozessgestaltung
36. VDI-2222(Bl.1): VDI-Richtlinie 2222 (1997) Konstruktionsmethodik—Methodisches Entwickeln von Lösungsprinzipien, VDI-Gesellschaft Produkt- und Prozessgestaltung
37. Aufreiter M (2008) Informationsvisualisierung und Navigation im Semantic Web: eine Analyse bestehender Visualisierungstechniken im HInblick auf Eignung für das Semantic Web, Hagenberg
38. Fluit C, Sabou M, Harmelen F (2008) Ontology-based Information Visualization: towards semantic web applications
39. Bügel O, Laufs U (2008) Einsatz innovativer Informations- und Kommunikationstechnologien. Fokus Technologie. Chancen erkennen, Leistungen entwickeln

# Chapter 4
# Strategic Foresight Methodology to Identifying Business Opportunities and Technology Trends

**David Güemes-Castorena and Juan L. Amezcua-Martínez**

**Abstract** This study aims to synthesize a model, based on previously developed methods by using control systems logic and serving as reference for systematic planning and strategic foresight in identifying technology trends and business opportunities. This model serves as a decision support system within a changing environment, allowing early detection of important business trends. In the first phase, we set the goal of what will be prospected. In the second phase, resources and capabilities are defined. The third phase defines the direction based on prior information. The first control loop is made up, serving as a fast-track feedback making it adaptable to the changing environment. A fourth phase serves to define action plans and measures to achieve the objectives; this stage feeds back to keep the system updated on the progress of measures and actions taken. This model is expected to respond to the following strategic foresight basic questions: what is being sought, where the technologies are, what direction should be taken according to the current environment, and how to get there.

## 4.1 Introduction

It is now common to find various methodologies that serve to give an approximation of the trend or the way a certain element will take in the near future. In a fast changing environment, where the globalization, the telecommunications, information, and most of the changes are exponentially increasing, companies

D. Güemes-Castorena (✉) · J. L. Amezcua-Martínez
Tecnológico de Monterrey, Campus Monterrey, School of Engineering,
64849 Monterrey, NL, Mexico
e-mail: guemes@itesm.mx

D. Cetindamar et al. (eds.), *Strategic Planning Decisions in the High Tech Industry*, DOI: 10.1007/978-1-4471-4887-6_4,

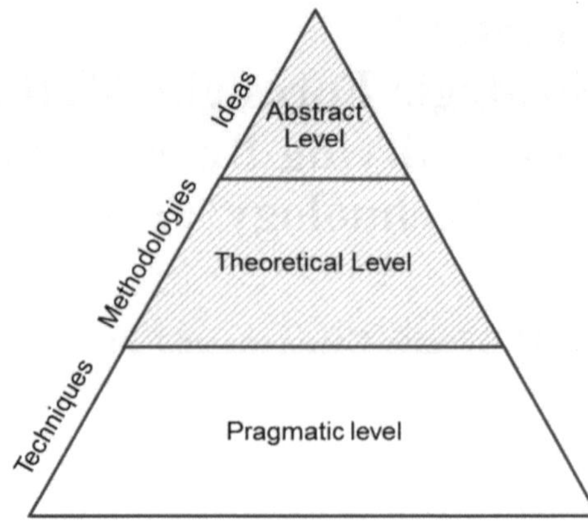


**Fig. 4.1** Representation of the abstract and theoretical levels (*shaded area*), the aim in this study

need to have the necessary people, skills, and tools to adapt to these quick changes in order to obtain or develop technologies and hence, innovate—or die.

It is not enough for a company to depend on the always-used group of methodologies that make a lineal foresight—or a trend analysis. It is necessary for companies to clearly define the road of their strategic actions according to the changing markets. Companies must be able to perform on time and with accurate proportion, to the inherent changes in an evolution toward an uncertain future; so in this way it reaches this objective by innovative means to apply innovation to achieve and construct a desired future.

This study aims to throw some light at the abstract and theoretical levels, and it is not intended to reach deeper levels—pragmatic, for example—yet (see Fig. 4.1) although the recommended tools can be used in the proposed order. Dealing with ideas and concepts as well as methodologies allows directors to shape directions to their organizations. The proposed methodology takes into consideration the recommended best practices that are presented in a logical framework. Right now we look forward to clarify the link between the control systems logic and the technological foresight methodologies.

Here, we propose a model that attempts to attach a set of methodologies that would enable an iterative foresight process based on the logic of the control systems. A *control design*, according to Goodwin [1], "aims to achieve a desired level of performance in the face of uncertainty". In a typical control system, there is a set point, which we call stage 1; there is a process (that we represent by stages 2 and 3), and there is a feedback loop (which will be stage 4). The conceptual Foresight Control System (FCS) has four main stages: (1) definition of goals, (2) delimitation/acknowledgement of resources, (3) immediate action or target definition, and (4) environmental evaluation, as illustrated in Fig. 4.2. We refer to the control systems theory in order to develop a foresight methodology to identify business opportunities since the four stages are analogous and there are always changes that need to be taken into account in the process. It must be recalled that the tools presented here are not the only tools that can be used for the proposed foresight process, but they are being considered as a good first approach.

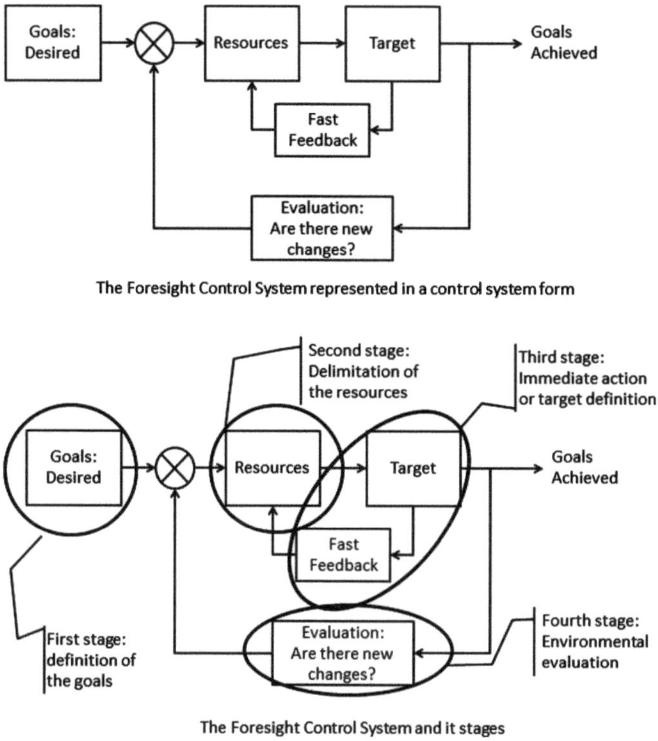


**Fig. 4.2** Foresight control system (*FCS*) and its stages for analysis

It is worth mentioning that an advantage of a feedback control system is that the system takes into account unforeseen changes that are slow as well as fast to detect and are fed into the process to determine a new set point—or direction.

## 4.2 Methodology

In this section, the stages of the proposed model are described, as well as the recommended tools for its implementation, based on the control system design body of knowledge and relating it to the needs for a systematic foresight process.

### *4.2.1 Stage 1: Definition of Goals and Vision*

This phase seeks to establish the aspects that define the system that will be studied, according to the interests of each organization or company, as well as the methods that can help as shown in Fig. 4.3. The definition of goals and vision is outside the

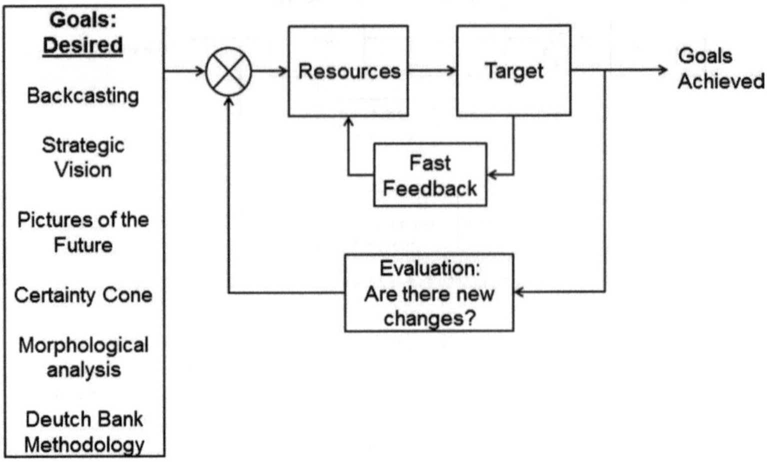


**Fig. 4.3** Recommended methods for stage 1

feedback loops. This is because the goals may change with time, or they may be different for each company, or for each project or area that will be prospected. This stage "sets the point" that will be maintained along the defined time interval, this being the life cycle of a product, or trend. For this purpose, the methodologies of this stage serve to outline the company's main objectives and the plan to achieve them. It is worth mentioning that the goals may change once the foresight exercise is started, because a trend that forces the goal to change may be identified; although this may happen, it occurs with very low frequency and therefore the assumption that it will not change is maintained in this study. Next, some methodologies that may help to build the first stage are presented.

One of the first methodologies is the Backcasting [2] which is similar to the methodology of "Pictures of the Future" [3] used by the Siemens Technology Accelerator (STA). It is convenient to use the Backcasting from the beginning, especially when the future vision is not very clear. Backcasting is a method to develop scenarios and identify the pathway to reach them, by presenting a future scenario and connecting it to the present situation by an analysis of how it can be possible to create the future scenario. This methodology has great planning and retrospective power because it generates not only the goals of the system, but also defines roles and introduces compromises [4]. The Backcasting is a comprehensive methodology, but insufficient by itself to make a prospecting exercise. That is why the Backcasting is exploratory in nature but aims to be concise and concrete. Therefore, it may cover not just the first stage, but also the second and even the third stages. Any company that is well grounded must have a strategic vision. Hence it is important to define not only what aspects are required to be achieved, but also to define how the company sees itself in the near future in a pre-defined time frame.

To give more consistency and robustness to the gathered information, the foresight methodology can be continued by using the Certainty Cone suggested by

Saffo to limit the range the ideas—that were generated in the previous methodologies—can move, that is, the range of feasibility of these ideas [5]. The cone of certainty seeks to find the trend line within a certain range of feasible events which can lead to the most probable future [5].

Morphological analysis and relevance trees [6] help us to obtain, based on a defined objective, all requirements and aspects involved for this purpose. That is why this tool will allow us to go from the goal, in retrospect, to all factors and resources that are required. As we shall see later, in stage 2, an analysis of the necessary resources or available skills to achieve those futures will be included.

Additional to this, we can complement the benefits with the application of the methodology of Deutsche Bank [7], whose forecasting system is based on the separation of scenarios, opportunities according to a two-dimensional matrix: of Impact-Certainty. As shown in Fig. 4.3, all these methods can contribute to the stage 1, Goals definition.

## *4.2.2 Stage 2: Resources (Science and Technology)*

Stage 2 is characterized by its focus clearly on the resources (mainly science and technology) and capabilities of the enterprise or system as shown in Fig. 4.4.

The fact of defining the necessary resources—or available skills—will allow us to analyze how feasible are the goals that were stated at the beginning. Until this stage, the process continues by gathering relevant information. We look to establish what information is available, whether it is resources, capabilities, or historical information. As one of the early stages, creative methods at this stage of information gathering can originate new ideas and proposals.

One of the main tools in this stage is the Environmental Scanning and Monitoring (ES&M) [8], but in the resources stage (Stage 2) the scanning is done while the monitoring corresponds mainly to the feedback stage 4 as the control

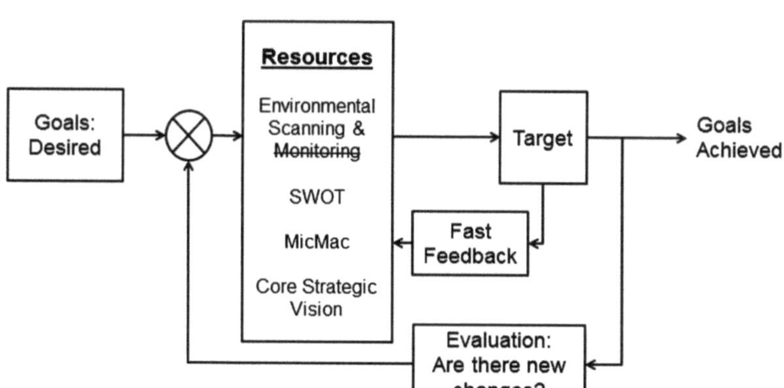


**Fig. 4.4** Recommended methods for stage 2

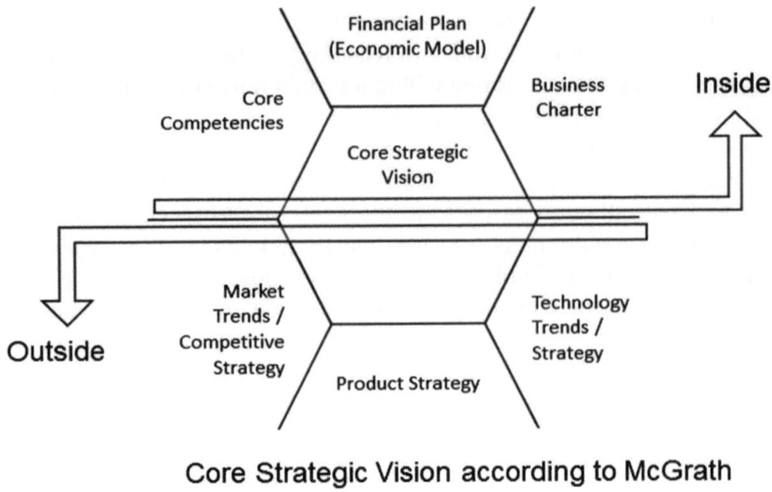


**Fig. 4.5** McGrath's core strategic vision

mechanism (this is the reason why the M is crossed). At the ES&M we should define those key variables that will be monitored later at the feedback stage.

Now we define the Core Strategic Vision, which according to McGrath [9] (see Fig. 4.5), will allow us to define a six-element scheme, as follows:

1. Core Competencies
2. Financial Plan
3. Business Charter
4. Market Trends/Competitive Strategy
5. Product Strategy
6. Technology Trends/Strategy

Then we can define the six strategic issues a company must consider both externally (Market Trends/Competitive Strategy, Product Strategy, Technology Trends/Strategy) and internally (Core Competencies, Financial Plan, Business Charter).

Added to this, the SWOT [10] analysis is one of the most used methods in this stage since, apart from defining the key competencies, it can serve as a starting point for discussions on straightforward actions that can detonate the strategic vision. The different points of view can initiate the discussion about the real opportunities and threats.

The SWOT improves the competitiveness aspects of a company so we can focus the efforts on innovation.

As the SWOT allows us to focus on strengths, minimize weaknesses, and gain the greatest advantage of the opportunities available, one could use the Pareto [11] method for weighting these opportunities.

As discussed above, the methodology of Deutsch Bank uses a matrix of Impact-Certainty; also, the SWOT analysis can develop an array of performance-importance to detect the key or critical factors [10].

All foresight analyses must take into account not only new information but also the one that has been generated over time. With this information, we will have the approximated path that the trend behavior will follow.

The historical data is considered at this stage, since this information serves as a base to get an approximation of how the system behaves. Therefore, every time the foresight control cycle is completed (Stage 3), it will be generating more historical data that can be fed back in this section.

Since this is an information-gathering stage, the amount of variables may grow, making the foresight exercise a very complex problem.

It is necessary to weigh those variables that are relevant for the system. With this in mind, the MICMAC [12] methodology is proposed because it is especially made for complex problems that involve a lot of variables. The MICMAC or Structural Analysis looks forward to represent the 'system' by highlighting key variables, which may influence the problem under study. It is important to ensure that only the important variables (internal and external) are being taken into account.

With methodologies like this, a complex system can be represented, highlighting those variables that affect (potentially) the study.

There may be some aspects that have not been considered at this stage of data gathering. That is why the creativity methods [13] provide the flexibility to adapt to new information and new insights that have not been described previously.

### *4.2.3 Stage 3: Target Direction Definition*

Defining the target direction is the most critical stage of the process; this is the reason for involving a greater number of methodologies. It is a stage of analysis and synthesis, where all the information previously obtained, interweave to define the results of the foresight.

Therefore, a wider set of methodologies is required as the result of analysis and synthesis of the previous information that will generate the trend and the foresight of the goals created at stage 1.

At this moment, it is very important to have previously defined time frames during which the foresight will be valid. For this we can refer to the recommendations for the foresight from Saffo [5].

If we start with the construction of scenarios, we can begin to generate situations with feasible and achievable projections, which can be mapped later in the cone of certitude [5].

Prior to that, the methodology of "Critical and Key Technology Study" [9] can give us a list of technologies that may be made from another list developed at an

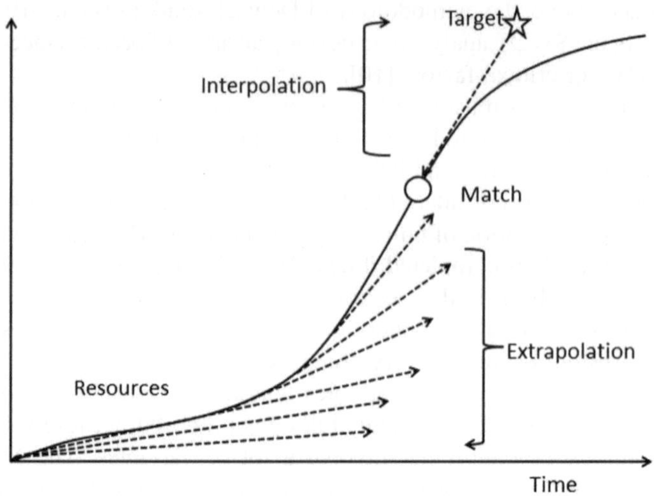


**Fig. 4.6** The match between the extrapolation and interpolation analysis

earlier stage (previous studies of foresight, patent analysis, bibliometrics, and so on.)

Now, based on clear and concise information, we can retake the cone of certitude tool to redefine the boundaries of the foresight study and have a more accurate result.

Along with the cone of certainty, we can combine the methodology of System Dynamics [14]. This tool searches the direction in which the system will evolve, by modeling it.

The analysis of trendinter-extrapolation (TI&E) [15] has the function of measuring changes in the different sectors to define the trajectory of the trend (extrapolating) and not only that, but also to match (interpolating) with the objectives and goals defined at the first stage as shown in Fig. 4.6.

This is the essence of the proposed magnitude and direction vector (an element of extrapolation in the fast feedback loop), which measures both the strength and speed of change (slope) as the direction in which they unfold.

The Technology Roadmapping (TRM) [16] is a tool which has graphical representations and is widely used to develop work plans. This tool strengthens the connection between objectives and resources, together with the Backcasting.

The TRM also serves as a support to the magnitude and direction matrix, used for the extrapolation at the Fast Feedback Control Loop (FFCL).

Also, the cross-impact analysis [17] is a tool that can be used to find relationships between these variables and the probability of how much both affect each other (Fig. 4.7).

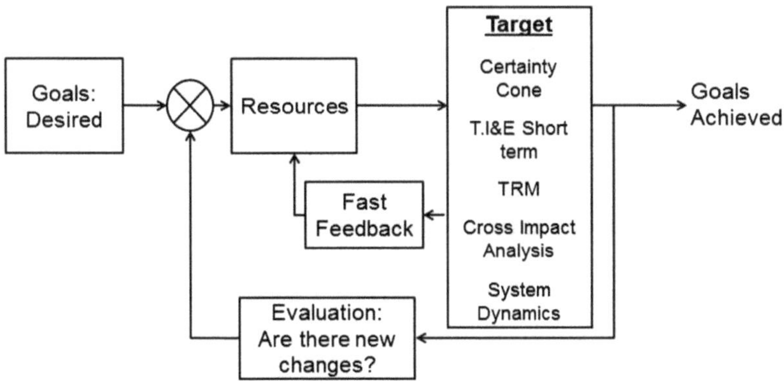


**Fig. 4.7** Recommended methods for stage 3—target definition

### *4.2.4 Feedback*

The importance of the feedback (shown in Fig. 4.8) is mainly to provide new and opportune information into stage 2. This will take corrective action if the change rate is very fast.

Previously defined, the time frame will set the rate at which the internal control cycle will perform. Also, another interval must be defined for the fourth stage, the outside control loop. This interval must be larger than the one of the internal loop.

For this stage, the ES&M methodology [8] is proposed, focusing mainly on "monitoring" the current conditions of the variables that were selected at the scanning process in stage 2. This action is relegated to stage 2. It must be recalled that new variables that may affect the foresight system can be introduced.

This completes the first control feedback loop of the FCS.

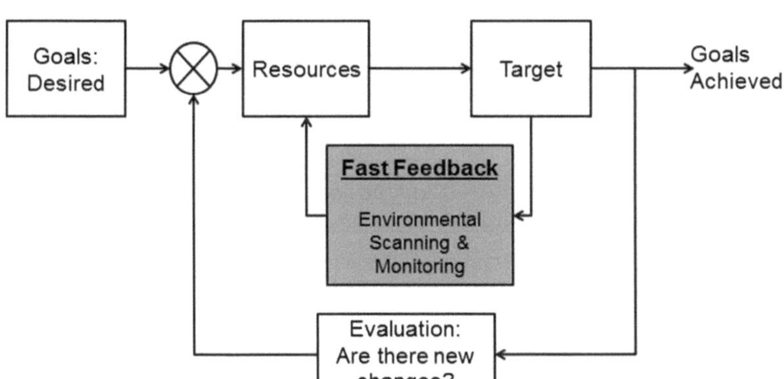


**Fig. 4.8** Environmental scanning and monitoring is the appropriate method for the fast feedback

### 4.2.5 *Stage 4: Feedback Evaluation: Global Strategy Management and Control*

This is the control stage at the strategic level. As in the fast control stage, the methodology ES&M. is used in the same way but with a more comprehensive and global approach.

It also uses the interpolation methodology to match the goals as the system evolves along the S-curve.

As in the Feedback stage, the FFCL uses the ES&M methodology (see Fig. 4.9) with a stronger focus on the monitoring (this is the reason why the S is crossed). This methodology allows us to have an environment's perception. This is also used at this stage with the same purpose but with a wider time frame than that one of the FFCL.

It is important to verify that the goals are still feasible because the time spans are wider. Therefore the long term TI&E, helps to:

1. Verify through interpolation, if you are following the right path to achieve these goals,
2. To review whether the goals have changed according to market conditions; or
3. Whether it is feasible to achieve the goals; or
4. Alternatively, the need to modify them.

Now, how this methodology is applied in the evolution of an S-curve of the trend during a certain time is shown in Fig. 4.10.

Having defined the objectives of the company by the methodologies as Backcasting, Morphological Analysis, and Relevance Tree (Point D, Stage 1), we can proceed to assess the situation and the available resources (Point A, stage 2). This second phase will include tools that allow us to evaluate the company both

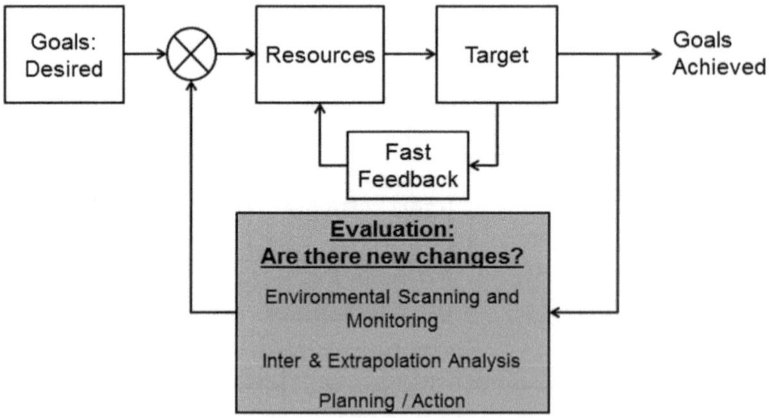


**Fig. 4.9** Recommended methods for the evaluation stage

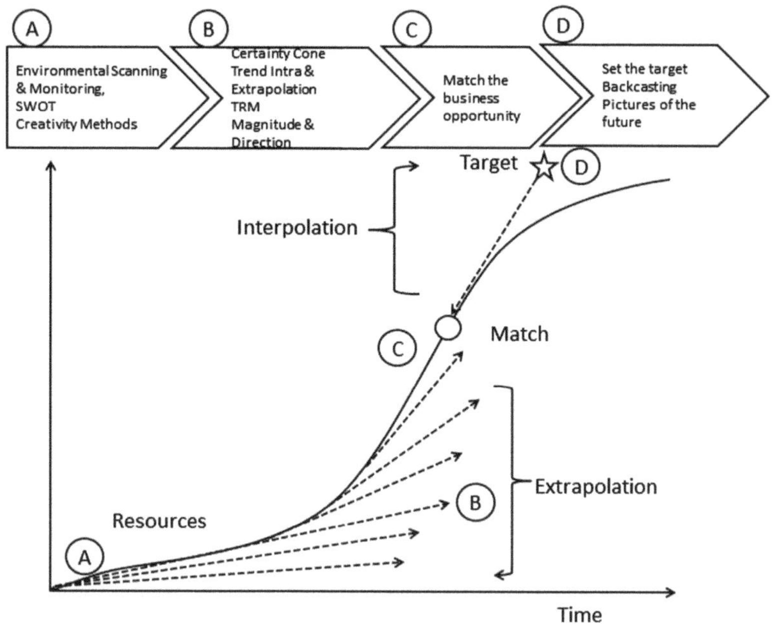


**Fig. 4.10** The foresight methodology in contrast with the S-curve

internally and externally (using SWOT and Core Strategic Vision). It will take into account the historical information and will perform the Environmental Scanning (ES&M), defining the critical system variables.

Based on the previously gathered information and monitored critical variables [18], we can start the extrapolation process, from the present to the future (point A to point C) and the interpolation, from a longer term future to the present (point D to point C). This is the obtained information from stage 3 (points B and C), as shown in Fig. 4.10.

As time passes, and based on a previous defined foresight time frame, point C will be traveling the S-curve from the initial point A, to the end point in D.

Returning to the control system, it is possible to identify how these points along the S-curve and the different stages of the FCS match (Fig. 4.11).

The control loops are defined as an iterative process that will complete monitoring feedback loops according to previously defined time frames.

## 4.3 Conclusions and Contributions

We aimed at applying control systems for technological foresight to identify an appropriate methodology that allows managers to get an objective and accurate as possible foresight or business opportunity detection.

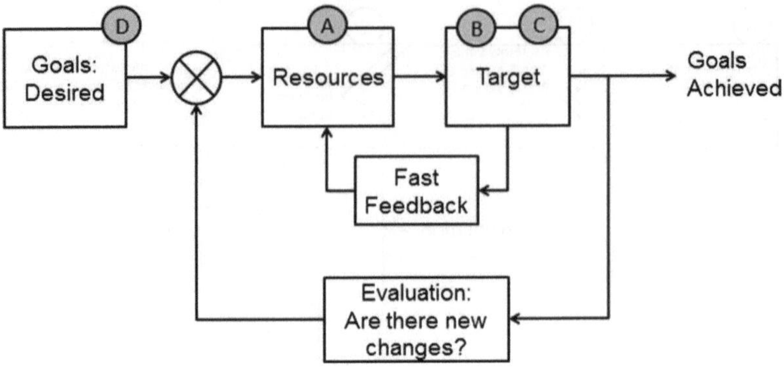


**Fig. 4.11** The foresight control system and the S-curve activities

When managers want to model complex systems which involve variables that are not yet known, say social variables, market behavior, etc., we realize that unlike the control systems used in industry, the methodologies here are not exclusive to a specific stage and therefore they may extend beyond one stage.

This study can be taken as a first attempt to develop a methodology that can parameterize the S-curve, so it is possible to have a more accurate approximation of the global trends. Also, the methodology seeks to systematize the existing procedures, and provides a deeper and more objective analysis and based on it—the analysis—synthesize a trend.

Through the control loops, the foresight system has the necessary robustness to keep itself updated and sensible to changes in the environment variables (internal or external) and the managers can make better decisions with better information[1].

## References

1. Goodwin GC, Graebe SF, Salgado ME (2000) Control design system. Valparaiso, Chile
2. European Commission-JCR-IPTS (2005) Backcasting. http://forlearn.jrc.ec.europa.eu/guide/4_methodology/meth_backcasting.htm. Accessed 2009
3. Theis D (2005) Pictures of the future. Siemens Jass
4. Robinson JB (1982) Energy backcasting: a proposed method of policy analysis. Energy Policy 10(4):8
5. Saffo P (2007) Six rules for effective forecasting. Harvard Bus Rev 11
6. European Commission-JCR-IPTS (2005) Morphological analysis and relevance trees http://forlearn.jrc.ec.europa.eu/guide/4_methodology/meth_morpho-analysis.htm. Accessed 2009
7. Schneider S (2007) The future of foresight in innovation and competitiveness. Deutsche Bank Research, Germany

[1] An earlier version of this paper was published in the Proceedings of PICMET '10, Technology Management for Global Economic Growth (Eds: Dundar F. Kocaoglu, Timothy R. Anderson, Tugrul U. Daim, Antonie Jetter, Charles M. Weber), July 18-22, Phuket, Thailand

8. European Commission-JCR-IPTS (2005) Environmental scanning and monitoring http://forlearn.jrc.ec.europa.eu/guide/4_methodology/meth_environmental-scanning.htm. Accessed 2009
9. McGrath ME (1995) Strategy requires vision. Product strategy for high-technology companies, 2nd ed. Two Penn Plaza, New York
10. European Commission-JCR-IPTS (2005) SWOT analysis http://forlearn.jrc.ec.europa.eu/guide/4_methodology/meth_swot-analysis.htm. Accessed 2009
11. Juran. (1989) Quality improvement tools. Juran Institute, Southbury
12. European Commission-JCR-IPTS (2005) Structural analysis http://forlearn.jrc.ec.europa.eu/guide/4_methodology/meth_structural-analysis.htm. Accessed 2009
13. European Commission-JCR-IPTS (2005) Creativity methods http://forlearn.jrc.ec.europa.eu/guide/4_methodology/meth_creativity.htm
14. European Commission-JCR-IPTS (2005) System dynamics http://forlearn.jrc.ec.europa.eu/guide/4_methodology/meth_systems-dynamics.htm. Accessed 2009
15. European Commission-JCR-IPTS (2005) Trend inter and extrapolation http://forlearn.jrc.ec.europa.eu/guide/4_methodology/meth_trend-extrapolation.htm. Accessed 2009
16. European Commission-JCR-IPTS (2005) Technology roadmapping http://forlearn.jrc.ec.europa.eu/guide/4_methodology/meth_roadmapping.htm. Accessed 2009
17. European Commission-JCR-IPTS (2005) Cross impact analysis http://forlearn.jrc.ec.europa.eu/guide/4_methodology/meth_cross-impact-analysis.htm. Accessed 2009
18. European Commission-JCR-IPTS (2005) Critical and key technology study (2005) http://forlearn.jrc.ec.europa.eu/guide/4_methodology/meth_critical-technology.htm. Accessed 2009

# Chapter 5
# Strategic Planning for Research-Intensive Institutions: An Approach Towards Identifying the Most Attractive Path for Technology Development

**Antonino Ardilio, Joachim Warschat and Dieter Spath**

**Abstract** The transfer of technology knowledge from research institutions into industry is a quite demanding task and often condemned to failure. In many cases, the main reason is the inability of research institutions to capture market needs and market requirements as well as consider them in their future technology development. But how to outfit research institutions or research-intensive enterprises with the rules of the free market economy and how to transform the identified actual and future market needs into the strategic planning for the technology? This chapter introduces a method for the identification of actual and future market needs and their transformation into the strategic technology planning using graph theoretical approaches.

## 5.1 Introduction

The intensification of competition, as a consequence of advancing globalization, not only affects enterprises increasingly, but also research facilities, as developers and providers of innovative technologies are advancing and are faced with an increasingly competitive environment [1–3].

A. Ardilio (✉) · J. Warschat · D. Spath
Fraunhofer-Institut für Arbeitswirtschaft und Organisation IAO,
Nobelstr 12, 70569 Stuttgart, Germany
e-mail: antonino.ardilio@iao.fraunhofer.de

J. Warschat
e-mail: joachim.warschat@iao.fraunhofer.de

D. Spath
e-mail: dieter.spath@iao.fraunhofer.de

D. Cetindamar et al. (eds.), *Strategic Planning Decisions in the High Tech Industry*, DOI: 10.1007/978-1-4471-4887-6_5,

Also, the funding of R&D activities is characterized by this change. Although the statistical data represented in the R&D data report of 2010 by the "Stiftungsverband für die Deutsche Wirtschaft" shows a stable relation between funding by the government and funding by the economic sector with respect to the financing structures of institutions for joint research (IJR). IJRs consist of organizations for basic research (as the Max-Planck-Gesellschaft) and applied research (as the Fraunhofer Gesellschaft) for years [4].

At the same time the report reveals that the state funding of R&D in the economic sector (namely enterprises and joint research institutions) has been decreasing continuously for the past two and a half decades [4] while simultaneously increasing the total R&D expenses of the government.

On the other hand the external efforts of the (economic) sector has risen considerably since the mid-1990s, even disproportionately compared to the internal R&D efforts (see Fig. 5.1) [4]. This aspect requires a more differentiating view.

More and more of state funding is flowing into public research (state research institutions, academies/universities) and into basic research (e.g., core/basic funding of Max Planck Gesellschaft), whereas the core funding of the applied research (like the Fraunhofer Gesellschaft) has developed disproportionately compared to basic research [5].

However, corporations are engaging preferably in applied research and in experimental development. The research contracts of companies which are awarded externally fluctuate even quite widely between different branches. Regarding manufacturing as an example of the business sector, their increasing effort in R&D is portioned as 20 % to academies, respectively, to state research institutes or even abroad, and 30 % to affiliated companies respectively to unaffiliated ones and IJR [4].

Therefore, it can be concluded that IJR, especially the ones engaged in applied research like Fraunhofer Gesellschaft, vie for a 30 % share of external research

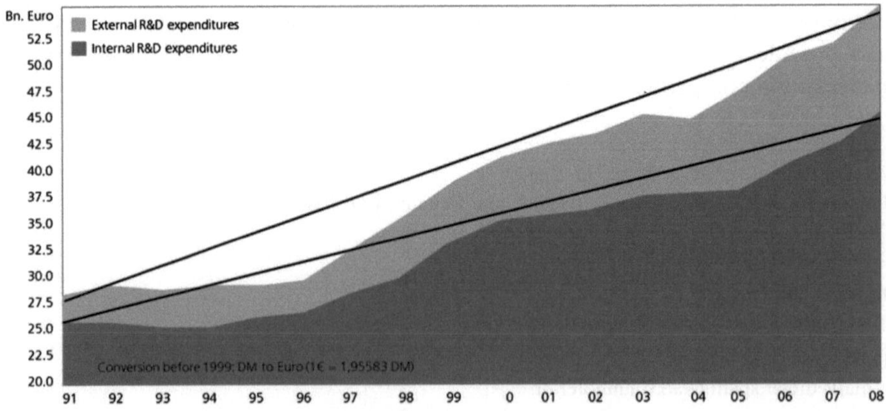


**Fig. 5.1** R&D expenses of the business sector

contracts from the business sector and thus have to become more and more attractive for it.

Another interesting aspect of financing of R&D is that a considerable part of the funding of IJR has been flowing in from abroad for several years [4]. Given this aspect and the fact that more and more IJR establish branches abroad, also R&D activities are becoming increasingly globalized [6].

In order to secure its financing under these conditions, a research institution must be and stay attractive for collaborations with industry [7–10]. Therefore it is inevitable for research institutions to recognize that they are subjected to the laws of free markets and hence focus more on commercializing its technology into the industry.

This requirement is emphasized by the following facts:

- Because of the intensification of the competition in technology and the competitive situation in consequence of globalization, public funding institutions are also focused increasingly on commercialization of research results, whereby technology has to be developed even increasingly orientated at customers demand and benefit.
  - "Which knowledge has to be produced in order to attract public funding in the future?"
- The progress in the information and communication technologies (ICT), as drivers of globalization, accelerates the diffusion of (technological) knowledge, while at the same time its half-life period decreases continuously. Hence, new requirements of producing and securing knowledge arise.
  - "Which knowledge has to be developed in order to address third-party funds successfully?"
- The rising external efforts in R&D from the business sector become an even more and more indispensable part of financing of IJR by third-parties (from the private sector), and will play a more and more important role in the growth and survival of (sate/public) research facilities.
  - "How to get more attractive to industry for cooperation?"

In order to keep the risks of development as low as possible for both technology user and technology developer, knowledge transfer between research facilities (or commercial technology developer) and corporations is inevitable [11]. Additionally, technology developer has to improve their often inadequate market knowledge, its requirements (namely in terms of customer needs) and especially research facilities need to formulate an explicit strategy for technology development and thereby be able to react appropriately to the changed (and still changing) framework conditions. In the past, the latter were in a comfortable position because of their established primacy in the research landscape, so there was no need for technology development based on a strategy [12].

There are several approaches existing within the framework of strategic technology planning. The aim of those methods is to gather a possibly complete picture of the technological environment for strategic decision making by means of future technology development [13–18].

In opposition to the approaches described in the literature so far, the presented method is supposed to support primarily technology providers with a scope on applied research to identify, take up, and assess the evident but also the latent needs of the technology market, and to use these findings—based on the most relevant markets—for the future development of the technology.

## 5.2 Approach

The approach described in this chapter aims at answering the question concerning the technology development strategy: "Which parameters of performance have to be considered in the future technology development in order to meet the requirements of the most attractive actual and future applications?"

The general approach therefore is presented by Ardilio [11]. Based on the above-mentioned requirements and existing process models of technology management [19, 20] a model architecture was developed (see Fig. 5.2). In this respect, from the functional profile of this technology, its technology market and its competition, the potential of a technology is conducted, which is then transferred to the strategic technology planning.

Based on this model architecture, the following four phases for the strategic technology planning for technology developer can be derived.

1. **Phase: Technology analysis:**

Within this phase, the Technology will be translated to its performed functions. Besides the decomposition in its function(s), the attributes[1] of the technology will also be mapped.

2. **Phase: Technology competitive analysis:**

Based on the above determined functions and attributes profile, a Technology competitive analysis will be implemented. Besides the future developments of existing technologies new technologies will also be identified.

[1] Attributes are physical and cognitive characteristics of a function and can be conducted from the functions of a Technology. Attributes map both, performance parameter of the Technology and requirements of the market.

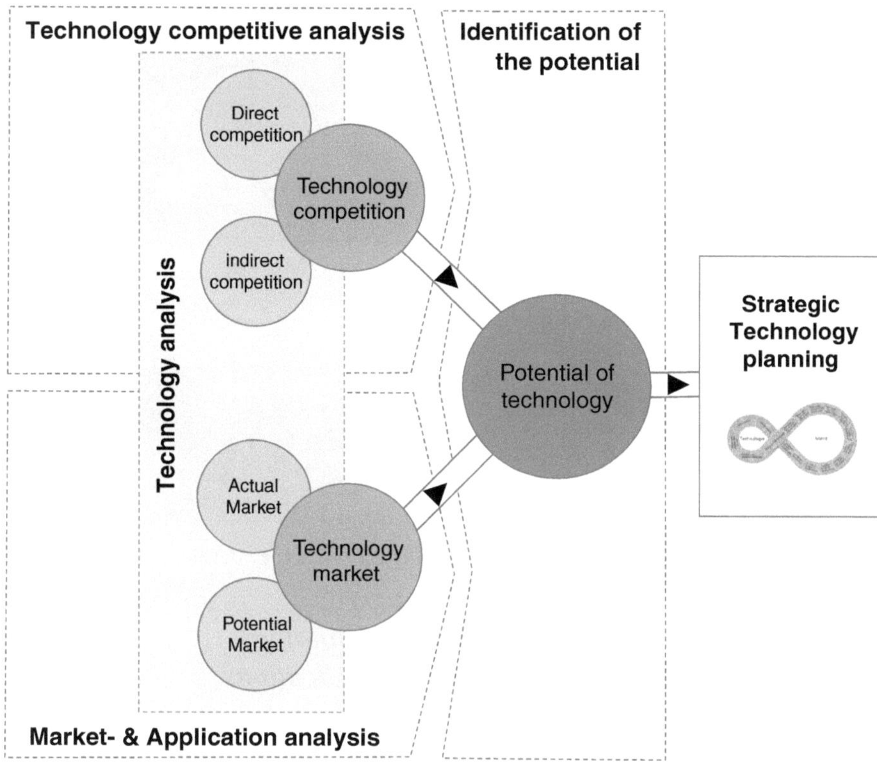


**Fig. 5.2** Architecture of the approach for strategic technology

3. **Phase: Application analysis:**

Within this phase new markets for the Technology will be identified. Besides already the existing and not yet addressed markets new markets will also be explored. For the found markets, single requirements will be determined.

4. **Phase: Technology marketing and strategic technology planning:**

Concerning the requirements and market position and competitive situation, the above found application will be transferred into a graph layout. Hence, directions of impact for the technology development will be determined and discussed.

In order to ensure a continuous process (as Ansoff's proclaims in his planning model), a cyclic approach was chosen to describe the method [21–23]. Figure 5.3 charts the approach. The Mobius-like Cycle is derived from the two poles "technology" and "market" (Fig. 5.3).

In the following the phases will be described and discussed in detail.

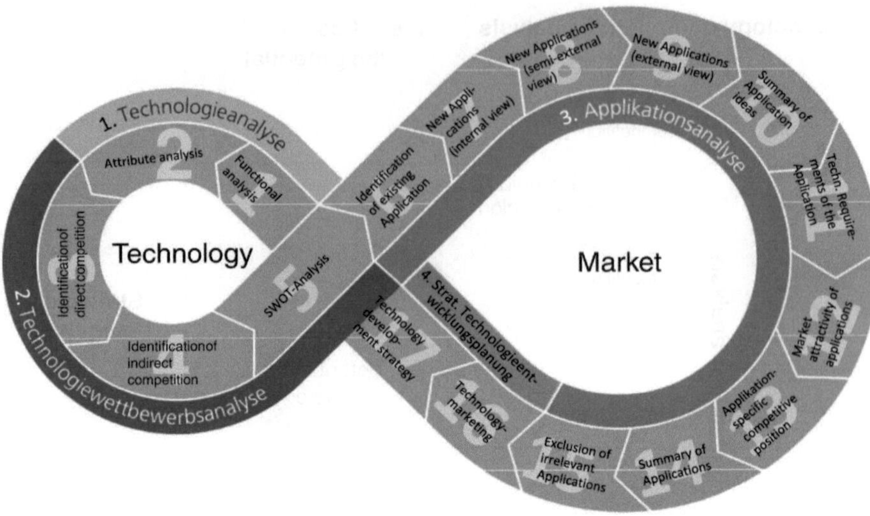


**Fig. 5.3** Approach for strategic technology planning for technology developer

## *5.2.1 Functional Description of the Technology and Identification of Potential Applications*

In order to be able to discover the relevant markets for technologies, current and potential applications for the technology have to be identified. Current applications in this context means applications which are already distributed in the market of the considered technology, while potential applications either are addressed by a competitive technology or are not existing in the markets at all yet.

The identification of applications for a technology is either orientated on the methods of idea generation or can be supported by intelligent patent research. In both cases, the functional profile of the technology plays a major role [24].

This profile will be mapped to different branches due to patent search or a creativity session based on branch lists (like NACE-Code etc.). If the provided function of a technology appears in patents, the associated IPC-Code gives a hint on the application field. The detailed description and methods of the identification of applications through creativity session and patent search can be found in Spath/Ardilio [11, 25].

From the discovered applications, those that were deferred for further consideration can already be addressed by the technology and thereby no further development of the technology is needed. Also, applications whose technological implementation is highly unlikely within the time period under consideration, were excluded from further consideration.

Outputs of this phase are potential applications for the technology which cannot be addressed yet by the technology due to performance or cost reasons.

In the next step the requirements are analyzed for every application and—according to the state of knowledge regarding the application—quantified. Since the majority of the numerous applications are initial ideas, an accurate specification of the requirements is often limited. Furthermore, it makes sense to specify qualitative intervals for the requirements in this phase (e.g., small, middle, big) or quantitative intervals (e.g., <1; 1–5; >5 kWh). The intervals have to represent the performance of the requirement in a reasonable scaling, which do not necessarily have to proceed in a linear direction. The analysis of the requirements of the applications takes place in a workshop. Here, the requirements of the applications for the technology should be evaluated and classified. Since the estimation and statements of the participants are affected by knowledge but also consumption and experience, a certain level of objectivity can be achieved by discussing the characteristics of the various criteria and searching for a collective result of all options. If applicable, the above-named economic organizations or networks of companies can be involved (see Table 5.3).

### 5.2.2 Identification of Market Indicators for the Identified Applications

For the strategic technology planning, besides the technical requirements of single application, the market attractiveness is also a relevant figure. Therefore—in connection with the identification of potential applications to be addressed—the attractiveness of the applications has to be determined by figures, respectively, by indicators, too.

Because the "market attractiveness" is a multidimensional variable of observation—especially when regarding technologies—its representation by only one single variable is not practicable [26]. By including further figures for the indication of market attractiveness, the effort also increases and at the same time the acceptance of the user of the measure sinks due to the increasing effort. Therefore, the set of figures should meet the requirements of providing the needed information as effectively and as efficiently as possible [27].

Following Diller [28] and others, the indicator of the market attractiveness of products is determined by the market size and the expected future growth. In many cases the expected margin is an additional market figure of relevance. Besides quantitative indices, qualitative indices can also be integrated, for example, the significance of the function within the application, which indicates whether the regarded function is a purchase crucial main function or a minor relevant additional function.

The market attractiveness of a technology has to be considered differently. It can be determined in advance, that in order to meet the abundance and diversity of the different applications, the inclusion of quantitative figures like "predicted unit sales" etc., is a task difficult to meet. Therefore, it is convenient to capture the market figures on a qualitative level.

Technologies show characteristics of products. Following the definition of Bullinger [19], technologies represent an approach to solving technical problems regarding/combined with business, organizational, and socio-political aspects. Additionally, regarding the fact that there is no recognizable or definable market for emergent technologies because of their novelty, a recourse/reference to the market indicators mentioned above seems obsolete.

Therefore, Lender suggests a consideration from two perspectives. Looking at the market orientation (corresponding to the application level in this article) and a given customer demand/need, a company creates the problem solving. In contrast, the technology orientation (corresponding to the technology level in this article) starts at the problem solving potential of the corporation and seeks for a new possible need [29].

Thus only then a technology is attractive for a technology customer, when he and/or his end customer benefits from the technology integration. Besides the direct customer benefit (technology customer), the indirect customer (end user) benefit also has to be considered with respect to the market attractiveness of the technology. Mollenhauer [30] calls this a chain of correlations, beginning with the end customers' buying motivation and proceeding to the success factors and the performance characteristics of the products and finally to the technology.

Thereby the attractiveness of a technology is described on two levels: on the technology market level and on the end customer market level. Figure 5.4 summarizes these indicators and market figures of the market attractiveness of a technology including the indicators described in detail in Sects. 5.2.2.1 and 5.2.2.2.

#### 5.2.2.1 Indicator of the Market Attractiveness at the Technology Level

The more applications a technology is able to address, the higher is its attractiveness. For the technology developer many addressable applications imply a high number of possible bilateral projects, licensing partners, or patent purchasers (see indicator $I_1$ in Fig. 5.4).

The identification of the number of potentially addressable applications (indicator $I_1$) for a specific stage of the technology development is realized by the comparison of the technology performance and the required application performance.

Thus $I_1$ results from the technology development path, because for a particular technology development stage those applications can be assigned, whose technological requirements profiles are met. $I_1$ is represented by a natural integer and only can become as big as the total number of addressable applications.

$I_1$ = number of addressable applications at given stage of technology development.

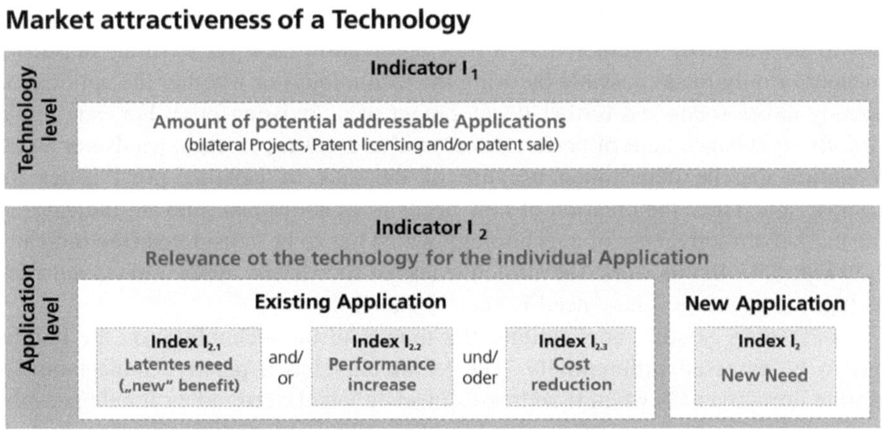


**Fig. 5.4** Indicators of the market attractiveness of a technology

Because this indicator results automatically from the stage of technology development, $I_1$ can be determined relatively simply and in a resource-saving manner.

### 5.2.2.2 Indicator of the Market Attractiveness at the Technology Level

The market attractiveness of a technology on the end user market level can qualitatively be determined by its enhancement of performance, the reduction of costs, and the potential to meet latent needs or fulfillment of demand.

The *performance increase* can be used as an indicator of the market attractiveness of a technology for existing applications (see indicator $I_{2.1}$ in Fig. 5.4). An enhancement of performance means in principle a higher performance of at least one attribute of a technology in comparison if the technology is already in use (e.g., "higher measuring accuracy"). Hence, a technology that provides higher performance when integrating into a product is more attractive.

An additional indicator of the market attractiveness of a technology for existing applications is the meeting of *latent needs* (see indicator $I_{2.2}$ in Fig. 5.4). Latent needs are regarded as the source of generating highly innovative ideas. The knowledge about these needs enables to identify new application fields for already existing technology, allow the exploring of potential application fields for emergent technologies, and can even trigger the exploration and development of new technologies [31].

*Cost reduction* is another indicator of the market attractiveness of a technology. The cost benefit of a technology is a benefit for technology—and end customer, especially concerning competing technology with the same performance (see indicator $I_{2.3}$ in Fig. 5.4).

For the determination of these indicators at the application level (indicator $I_2$) it has to be clarified, whether it is a new application (not yet existing in such a manner or only made possible by using the technology) or whether the application already exists (or is it a further development of an existing one). For *new applications* no enhancement of performance, neither meeting of latent needs nor a cost reduction can be determined because of the lack of existing possibilities for comparison. Thus, the creation of new needs is an additional relevant indicator of the market attractiveness of a technology which has to be considered (see indicator $I_2$ in Fig. 5.4). In this work, the attractiveness of such a new application is regarded as high, as it meets a new need in most instances.

Concerning existing applications, the impact of the technology on the market has to be assessed differentially. The technology has a positive impact on the market attractiveness when it addresses/meets a latent need, when it enhances the performance of the function provided by the technology, and/or when the costs of the application can be reduced.

The relevance of these three indicators are weighted differentially, because the addressing of latent needs has a higher influence on the market attractiveness of an application than the reduction of costs. In this work, the technology with the highest relevance is the one that addresses a latent need (50 %), followed by the one that enhances the performance of a function (30 %) and finally, the one that helps to reduce costs (20 %; see Table 5.1).

$$I_2 = \Sigma I_{2.i}\ (i = 1 - 4) \quad \text{when market already exists} \tag{5.1}$$

$$I_2 = 10 \quad \text{when market is new}$$

.

In principle the use of a technology in an application can exhibit all or at least several indicators simultaneously (e.g., enhancement of performance and cost reduction). The measurements of the indicators are conducted on a qualitative level and show a binary character (yes/no). After determining, the indicators are consolidated in the indicator $I_2$ ($I_2$ is the sum of the three indicator $I_1$–$I_4$).

For $I_2$ a scale ranging from 0 to 10 is chosen. $I_2 = 0$ means that none of the three indicators got a “yes”, $I_2 = 10$ means that all three indicators apply to the application. Table 5.1 shows an example for how the indicator $I_2$ can be deduced.

### *5.2.3 Determination of the Application-Specific Competitive Position*

The competitive position of a technology related to a specific application provides information about the degree of competition for the respective application and hence, it is a measure for the potential for use of the technology.

**Table 5.1** Plotting market attractiveness-indicator $I_2$ (example)

| | **New market** | **Existing market** | | | |
|---|---|---|---|---|---|
| | **$I_{2.0}$ new need** 10 Pt. | **$I_{2.1}$ new benefit** weight 50% 5 Pt. | **$I_{2.2}$ Performance increase** weight 30% 3 Pt. | **$I_{2.3}$ cost reduction** weight 20% 2 Pt. | **$I_2$** sum |
| **$App_1$** | 10 | 0 | 0 | 0 | **10** |
| **$App_2$** | 0 | 0 | 3 | 0 | **3** |
| **$App_3$** | 0 | 5 | 3 | 0 | **8** |
| … | … | | … | … | … |

Example

**New market:**

$App_1$: "Measure of dynamics of a signature as a security feature of credit cards"

no existing market yet → new market

**→ $I_2 = I_{2.0} = 10$**

**Existing market:**

$App_2$ : "Oscillation monitoring for machine tools"
existing market, enhancement of performance of function "tracking" (resolution)

**→ $I_2 = I_{2.2} = 5$**

$App_3$: "Supporting of GPS at shadowed areas"
existing market, latent need of existing market ("position detecting in shadowed areas") and enhancement of performance of function "positioning"

**→ $I_2 = I_{2.1} + I_{2.2} = 8$**

The incorporation of this indicator is relevant for answering the question addressed in this work, because the attractiveness of an application can be assessed more reliably based on the technology competition (in comparison to competing technologies).

A good competitive position of a technology increases the potential for its use within the regarded application. Because the competitive position of a technology related to a specific application influences the attractiveness of an application, it

**Table 5.2** Competitive position $I_3$

| $I_3$ | Addressing of application ... ..today.. | ..in the future.. |
|---|---|---|
| **0** | ..only by competing technology | ..only by competing technology |
| **1** | ..by own and competing technology | ..only by competing technology |
| **2** | ..by own and competing technology | .. by own and competing technology |
| **3** | .. by own and competing technology | ..only by own technology |
| **4** | .. only by own technology | .. only by own technology |

has to be incorporated for the determination of the future direction of development of the technology.

In order not to demand too much of the user for the determination of the indicator of the "competitive position related to a specific application" (competitive position), this qualitative indicator is classified in values between 0 and 4. The users have to clarify, whether the single application can be addressed exclusively by the own technology, exclusively by competing technologies, or something in between.

During the determination of the indicator, today's competitive situation is distinguished from the potential situation in the future. This takes account of the fact that also the competing technologies will be further developed in the future; an application not addressed yet by competing technologies in the future might be. Table 5.2 shows the individual evaluation levels of the indicator "competitive position".

## *5.2.4 Summarization of the Collected Information*

All identified applications, their technological requirements (based on the specification of the attributes), their indicator of market attractiveness (only indicator $I_2$, indicator $I_1$ results from data processing in a later stage), and the competitive position related to the specific application are merged in an application list (see example in Table 5.3).

The application list displays all the recorded and processed information and provides a basis for the strategic technology planning, which is introduced and discussed in detail in the procedural section.

The main result of this phase is a classification of all identified applications considering technical, competitive, and market-relevant criteria. An exemplary

**Table 5.3** Application list (example)

| Application | Technical requirements | | | Market requirements | | | | Sum |
|---|---|---|---|---|---|---|---|---|
| | | | | New Market | Existing Market | | | |
| | Attribute 1 | ... | Measure accurancy | | Latent need | Performance increase | Cost reduction | |
| | | | | 10 Pts | 5 Pts | 3 Pts | 2 Pts | |
| $A_1$ | 2 | .. | 3 | x | - | - | - | 10 |
| $A_2$ | | | | - | - | x | - | 5 |
| .. | | | | - | x | x | - | 8 |
| $A_x$ | | | | | | | | ... |

**Legend (e.g.):**
**1** can be met today
**2** can be met within 2 years
**3** can be met within 4 years
**4** can be met within 6 years

**Legend:**
"Easy-to answer"-questions:

1. Is it a new market or an existing on?

→ For existing Markets:
- Does the implementation of the technology meet a latent need?
- Does the implementation of the technology enhance the performance?
- Does the implementation of the technology reduce costs?

summary is shown in Table 5.3. Besides these direct results, one gains also the following indirect and not quantitatively measureable results:

- Sensitization of the technical educated participants with respect to market-orientated thinking.
- Sensitization of the employees with regard to an increased active and passive search for options for the institutions.

Finally, the identified application ideas, their requirements, and the competitive classification provide a holistic view of the requirement profile for the technology development.

## *5.2.5 Strategic Technology Planning*

For the strategic technology planning the information gathered so far concerning the market attractiveness and the technological competitive position (see Sects. 5.2.1.1 and 5.2.1.2) will be drafted in the further examination.

The strategic technology planning takes place in time-depending planning horizons:

- **Short-term strategic planning**
  Marginal optimization of the performance parameters, addressing market segments with a low growth regarding the performance.
- **Medium-term strategic planning**
  Medium optimization of the performance parameters, addressing market segments with a medium growth regarding the performance.

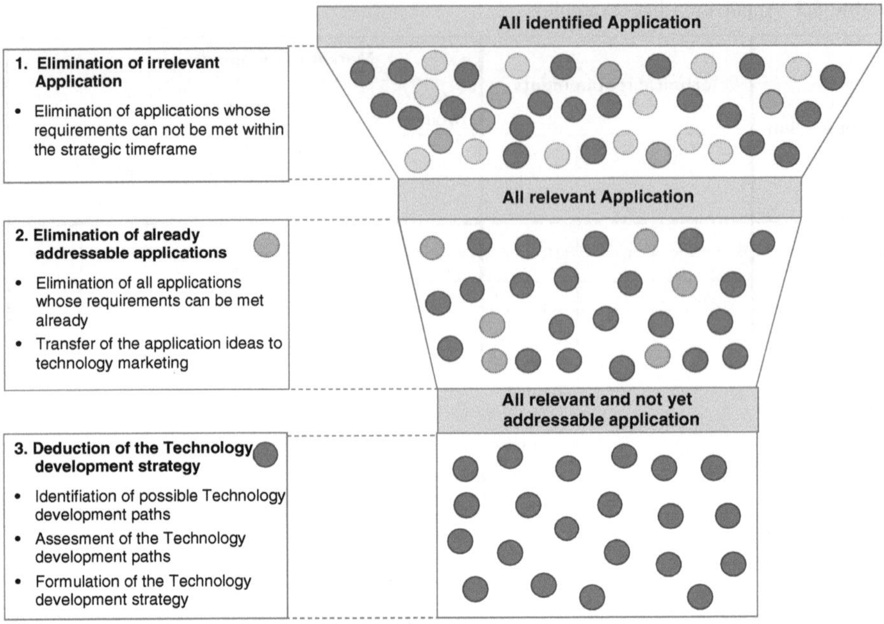


**Fig. 5.5** Tasks within the strategic technology planning

- **Long-term strategic planning**
  Significant optimization of the performance parameters, addressing market segments with a large growth regarding the performance.

In order to formulate the strategic technology planning, all possible technology development paths will be deduced from the found applications. These paths will then be prioritized concerning their attractiveness.

Within a technology development path the single phases of technology development will be stated in a chronological order. The starting point of each path is the actual performance of the technology (concerning all attributes). The ending point of each possible path derives from the application with the highest performance needed (concerning all attributes). Figure 5.5 describes the remaining tasks for the strategic technology planning [32].

### 5.2.5.1 Elimination of Irrelevant Applications

In order to identify the most attractive technology development path, first the irrelevant applications have to be eliminated (Fig. 5.5).

As stated already in Sect. 5.2.1 not all found applications can be considered in the strategic technology planning. All applications, whose technological implementation is highly unlikely within the time period under consideration were excluded from further consideration already. All applications, whose requirements already can be fulfilled by the technology, are not to be considered for this task, as well. Anyway, those applications should be transferred to technology marketing [11].

Applications with low market attractiveness should not be eliminated yet. Depending on the chosen technology development path they may be able to be addressed "by the way" without any additional development resources needed.

### 5.2.5.2 Technology Development Strategy

The aim of this task is to derive the strategic technology planning based on the requirement profiles of the most attractive applications. The required time resources for development of a technology depend strongly on the needed performance-profile for the applications. This performance profile correlates to the technological requirements (see Table 5.3).

The number behind an attribute in Table 5.3 states the resource-intensity of the technology developer to meet the requirements. The higher this number is, the more time-intensive is its development and the more long term is the time horizon for technology development.

Due to this fact, the applications can be brought into a temporal order concerning their development intensity; applications with low attributes can be addressed earlier than applications with high attributes.

In this work it is assumed that each of the independent attributes can be improved successively in chronological order only, i.e., just one attribute per technology generation is improved.

Hence, within all possible technology development options that path is to be identified, which offers the largest marketing chances and the lowest degree of competition on a long-term basis for the technology. Unfortunately, there are many possible paths, which make the selection difficult.

Concerning the attractiveness of the technology development path the following statement can be derived: A technology development path is attractive, if

- the intermediate technology development conditions address many applications along the technology development path ($I_1$).
- The attractiveness of the applications along the path is very high ($I_2$).
- The competitive position of the technology for the applications along the path is very strong ($I_3$).

To identify this path, methods of the graph theory come in use. Hereby, the path search problem can be solved mathematically by an acyclic arranged multigraph [33].

Hereby, the applications can be transferred into a graph. The connections of the applications (junctions) represent the possible technology development options (Fig. 5.5). The graph begins with applications, whose attributes can

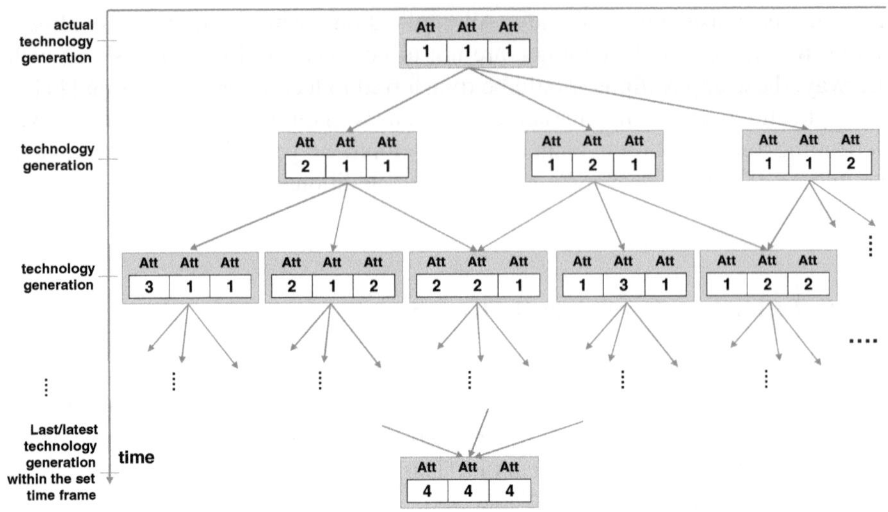


Fig. 5.6 Theoretically possible technology development options (acyclic arranged multigraph)

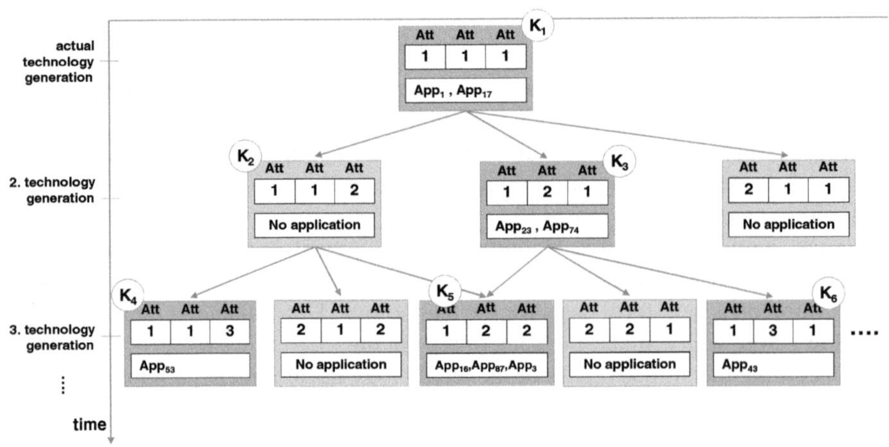


Fig. 5.7 Technology development options (example)

already be fulfilled today; this corresponds to those applications, whose attributes match to "1" (see Fig. 5.5). The tree ends with the applications which have the maximum requirement profile; here all attributes of the applications correspond to "4" (see Fig. 5.6).

First, all paths will be mapped by raising one of the attributes up. Within all possible paths in the following, only those paths will be considered, whose knots contain at least one application (e.g., $K_3$ in Fig. 5.7). Also, paths which have an

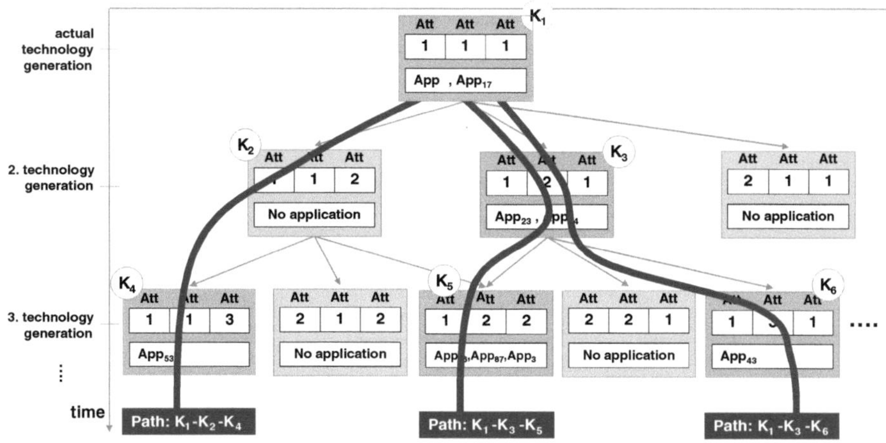


**Fig. 5.8** Technology development paths (example)

empty knot can be considered, if the following knots contain at least one application (e.g., $K_2$ in Fig. 5.7).

Based on the graphs, it is now possible for the user to derive and estimate the possible technology development options. The example in Fig. 5.8 points out the following technology development paths:

| | |
|---|---|
| $K_1$–$K_2$ –$K_4$ | With this technology development path altogether three applications can be addressed (up to the third technology generation), whereby the second technology generation does not include applications |
| $K_1$–$K_3$ –$K_5$ | With this technology development path altogether seven applications can be addressed (up to the third technology generation) |
| $K_1$–$K_3$ –$K_6$ | With this technology development path altogether five applications can be addressed (up to the third technology generation) |

As the attractiveness of a technology can be described by the amount of application it can address and concern the market attractiveness of the application (see Sect. 5.2.2), for example, even a path including three applications (with high market attractiveness) can be more attractive than the path including seven applications (with low market attractiveness).

In order to consider all relevant indicators in the strategic technology planning, the market attractiveness (Sect. 5.2.2) and the competition intensity (Sect. 5.2.3) are also included in the graph (see Fig. 5.9).

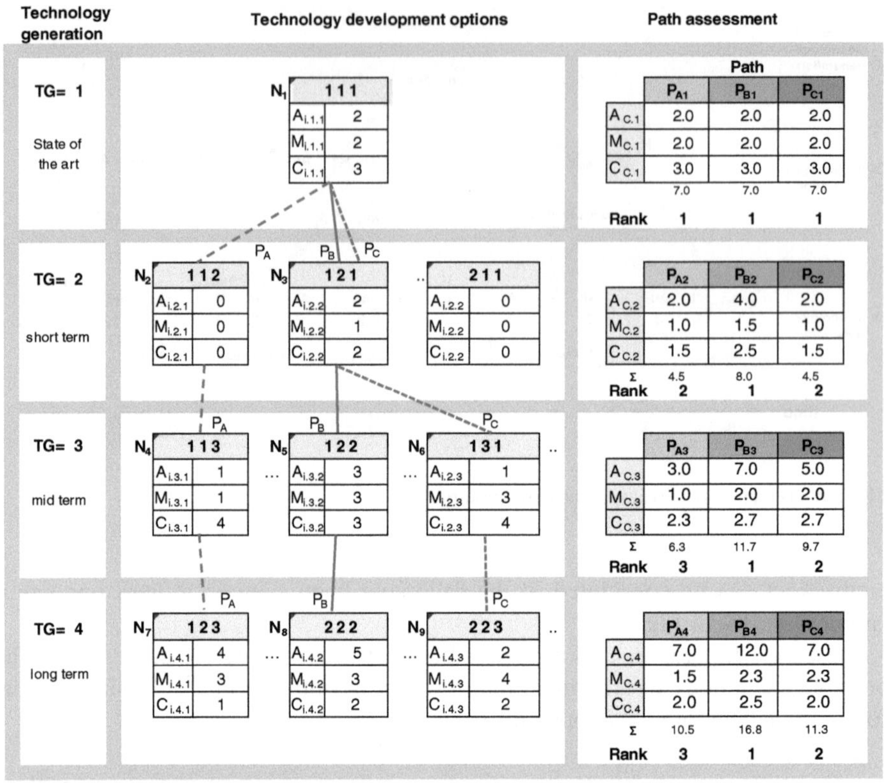


**Legend**

| | |
|---|---|
| TG : | Technology generation |
| $N_n$ : | Technology development option |
| $A_{i.x.y}$ : | Amount of Application which can be addressed with this technology development option (not cumulated along all TG) |
| $M_{i.x.y}$ : | Market attractiveness of applications which can be addressed with this technology development option (cumulated along all TG) |
| $C_{i.x.y}$ : | competition intensity of the Technology within the application which can be addressed along the path (cumulated along all TG) |

**Fig. 5.9** Assessed technology development options (example)

With all these numbers it is possible now to calculate the most attractive path along all possible ones. Hence, the attractiveness of the single development options (see column "Technology development options" in Fig. 5.9) within a technology generation will be determined (see column "path assessment" in Fig. 5.9). The following formulas are used:

$$\text{Cumulated number of applications:} \quad A_{c.n} = \Sigma\, A_{i.x.y} \text{ along the path} \tag{5.2}$$

$$\text{Average attractiveness of applications:} \quad M_{c.n} = \frac{\Sigma\, M_{i.x.y} \text{ along the path}}{TG} \tag{5.3}$$

$$\text{Average competition intensity of applications:} \quad C_{c.n} = \frac{\sigma\, C_{i.x.y} \text{ along the path}}{TG} \tag{5.4}$$

$A_{C.n}$, $M_{C.n}$ and $C_{C.n}$ gives an indication of the attractiveness of the single technology development options along a path.

For the assessment of the complete path, a utility value analysis (introduced by [34]) will be used, including all technology development options along the single paths. Based on the ranking of the technology development options within a technology generation the utility value analysis will be calculated (see Table 5.4).

**Table 5.4** Utility value analysis of the technology development paths (example)

| Path | TG = 2 (short-term) 20 %[a] | TG = 3 (mid-term) 50 %[a] | TG = 4 (long-term) 30 %[a] | Rank | |
|---|---|---|---|---|---|
| $P_A$ | 0.4 | 1.5 | 0.9 | 2.8 | |
| $P_B$ | 0.2 | 0.5 | 0.3 | 1 | ←Most attractive path is $P_B$ |
| $P_C$ | 0.4 | 1 | 0.6 | 2 | |

[a] Weighting of the utility value: Focus on midterm

In the example shown in Fig. 5.9 the technology development path $P_B$ is the most attractive. The amount and the market attractiveness of the addressable applications within this path is the highest and the degree of competition is the lowest in comparison to the other possible paths.

Therefore in this example, the technology development strategy can be formulated as follows:

- **Short-term strategic planning**
  For second Technology generation (TG = 2) attribute $Att_2$ should be improved.
- **Medium-term strategic planning**
  For the third Technology generation (TG = 3) attribute $Att_3$ should be improved.
- **Long-term strategic planning**
  For the third Technology generation (TG = 4) attribute $Att_1$ should be improved.

Consequently, from the technology development strategy, a plan of action for the implementation of the strategic issues should be elaborated and the achievement of the objectives should be controlled in specific intervals.

## 5.3 Conclusion

The need for an instrument for the support of the technology developer in their technology development planning is growing due to the increasing complexity and convergence of technology disciplines and the rising dynamics of new technology development.

The goal of Strategic planning for research institutions is to support the technology developer in the formulation of a technology development strategy (and in the marketing of current technology) in order to maximize the financial benefit of their research efforts.

The approach is based upon the identification and analysis of the competing technologies, competitors, and the actual and potential markets for a technology. The approach was divided into three main phases. In the technology analysis phase the current and potential competing technologies and the competitors are identified and described. In the application analysis phase the established and emerging markets are determined on an application level. In the potential investigation phase the applications are transferred into a graph, in order to find the most attractive technology development path. From this path, the technology development strategy can be formulated.

A central component of the approach is the application analysis and the strategy portfolio. With the application analysis the branch force-fitting method (through the use of the NACE code as structure) shows that a large and varying amount of application ideas could be generated.

It was shown that the technology development graph offers a clear visualization of the complex connections between current and/or potential technologies, competitors and markets and thus it was very useful for answering the question specified above.

In order to enable the technology provider to examine and/or to adapt its technology marketing and/or technology development strategy, it would be advisable to implement the technology strategy planning on a regular base.

The collection of the necessary data both for the initial take-up of the data and for its continuous extension and adaptation is resources intensive.

It is shown that this graph-based representation summarizes all relevant information for decision making concerning the technology development strategy and thus serves as practicable decision basis for strategic technology planning. However, the application of the approach within several institutions of the Fraunhofer Society showed, that even for technologies whose fields of deployment are still indistinct, the identification of potential applications and the formulation of the technology development strategy for a short-, mid-and long-term time frame was successful.

Scientifically the introduced approach must be widened to include also technology systems (consisting of more than one technology) and to the fact, that more than one attribute per technology generation can be improved.

## 5.4 Suggestions to Managers

Besides increasing the market share in existing markets, the identification of new markets—outside of those known and addressed by the company—is a way to increase turnover. However, new markets may need technological adaptions, which therefore need resources.

In order to successfully diversify into not-yet-known markets, the Manager should be able to answer the following questions:

1. What is my technology portfolio able to fulfill?
2. Who are the competitors within our technology field and which other competing technologies are existing?
3. What are the unique selling points of our technology portfolio?
4. Which markets require those unique selling points?
5. How attractive are those markets?
6. What are the requirements of the most attractive new markets?
7. Do we meet those requirements?

8a. If yes: What do we need to do to address them ("low hanging fruits")?
8b. If not: When can we address which of those relevant markets and how much efforts do we need to invest ("high hanging fruits")?

When addressing new markets which need further technology development, strategic thinking is a key issue for a sustainable business development. The following metaphor illustrates this subject: "Harvest the low hanging fruits but think about which path to climb the tree in order to get as much as possible from the higher hanging big apples".

## References


1. Mintzberg H (1994) The rise and fall of strategic planning. Prentice Hall, London
2. Frauenfelder P (2000) Strategisches management von technologie und innovation: tools und principles. Industrielle Organisation, Zürich
3. Mittag H (1985) Technologiemarketing die vermarktung von industriellem wissen unter besonderer berücksichtigung des einsatzes von lizenzen. Brockmeyer, Bochum
4. Sahlman K, Haapasalo H (2009) Perceptions of strategic management of technology in small high-tech enterprises. In: Kocaoglu DF (ed) Technology management in the age of fundamental change. Portland, Oregon
5. Abele T (2006) Verfahren für das technologie-roadmapping zur unterstützung des strategischen technologiemanagements. Jost-Jetter, Heimsheim
6. Friar JHM (1985) The emergence of technology strategy: a new dimension of strategic management. Technol Soc 7:143–178
7. Bullinger HJ (2006) Fokus innovation: kräfte bündeln—prozesse beschleunigen. Retrieved from http://deposit.ddb.de/cgi-bin/dokserv?id=2712793&prov=M&dok_var=1&dok_ext=htm. Hanser, München

8. Möhrle M (ed) (2008) Springer-11774/Dig. Serial. Technologie-roadmapping: zukunftsstrategien für technologieunternehmen (3., neu bearbeitete und erweiterte auflage.). Retrieved from http://dx.doi.org/10.1007/978-3-540-74755-0. Springer, Heidelberg
9. DeGeeter MJ (1998) Technology commercialization manual: strategy tactics and economics for business success. University of Illinois, Urbana
10. Hamel G (2006) The why, what, and how of management innovation. Harvard Bus Rev 84(2):72–84
11. Spath D, Ardilio A, Laib S (2009) The potential of emerging technologies: strategy-planning for technology-providers throughout an application-radar. In: Kocaoglu DF (ed) Technology management in the age of fundamental change. Portland, Oregon
12. Spath D, Bullinger H-J (eds) (2004) Forschungs-und technologiemanagement: potenziale nutzen—zukunft gestalten; [Festschrift zum 60. Geburtstag von Prof. Hans-Jörg Bullinger]. Retrieved from http://www.worldcat.org/oclc/645190401. Hanser, München
13. Lichtenthaler E (2000) Organisation der technology intelligence. Eine empirische untersuchung der trechnologiefrühaufklärung in technologieintensiven grossunternehmen. ETH, Zürich
14. Schöning S (2006) Potenzialbasierte bewertung neuer technologien. Retrieved from http://www.gbv.de/dms/hebis-darmstadt/toc/177396148.pdf. Shaker, Aachen
15. Savioz P (2004) Technology intelligence: concept design and implementation in technology-based SMEs. Palgrave Macmillan, Basingstoke
16. Zinser S (2000) Einevorgehensweise zur szenariobasierten frühnavigation im strategischen technologiemanagement. Jost-Jetter, Heimsheim
17. Gomeringer A (2007) Eine integrative, prognosebasierte vorgehensweise zur strategischen technologieplanung für produkte. Jost-Jetter, Heimsheim
18. Brink V Entwicklung konsistenter technologiestrategien mit der innovationsdatenbank. In: Vorausschau und Technologieplanung—3. Symposium für vorausschau und technologieplanung Heinz nixdorf institute, pp 239–261)
19. Bullinger H-J (1994) Einführung in das technologiemanagement: modelle, methoden, praxisbeispiele. Teubner, Stuttgart
20. Servatius H-G (1985) Methodik des strategischen technologie-managements. Erich Schmidt, Berlin
21. Mintzberg H, Quinn JB, Ghoshal S (1998) The strategy process (Rev. European). Retrieved from http://www.worldcat.org/oclc/37782979. Prentice Hall, London
22. Gälweiler A, Schwaninger M (1986) Unternehmensplanung: grundlagen und praxis (Neuausg./). Campus-Verlag, Frankfurt
23. Spath D, Dill C, Scharer M, Wolski A (2001) Vom markt zum markt: produktentstehung als zyklischer prozess. Retrieved from http://www.worldcat.org/oclc/76361076. Logis, Stuttgart
24. Spath D, Ardilio A, Warschat J (2011) Technologiemanagement: radar für erfolg, 1st edn. Log X. Retrieved from http://www.worldcat.org/oclc/767800381. Württ, Ludwigsburg
25. Ardilio A, Warschat J (2010) Win$^3$ a SME-customized approach towards a sustainable technology strategy. In: portland international conference on management of engineering and technology (PICMET). Phuket, Thailand
26. Krause O, Mertins K (2006) Performance management—eine stakeholder-nutzen-orientierte und geschäftsprozess-basierte methode. Deutscher Universitätsverlag, GWV Fachverlage GmbH, Wiesbaden
27. Savioz P Technology intelligence systems: practices and models for large, medium-sized and start-up companies. Int J Technol Intell Plann. Retrieved from http://www.inderscience.com/search/index.php?action=record&rec_id=11707&prevQuery=&ps=10&m=or, pp 360–379
28. Diller H, Bukhari I (1998) Marketingplanung. Vahlen, München
29. Gassmann O, Kobe C (2006) Management von innovation und risiko: quantensprünge in der entwicklung erfolgreich managen. Retrieved from http://dx.doi.org/10.1007/3-540-33755-5
30. Mollenhauer M, Remmerbach K-U (1989) Neue spielregeln des marktings: wie aktivieren wir die märkte für die nächste produktgeneration? In: Arthur D Little international (Hg.)—das management des geordneten wandels, pp 123–137

31. Kocaoglu DF (ed) 2009 Technology management in the age of fundamental change. Portland, Oregon
32. Spath D, Ardilio A, Warschat J (2010) Technologie/markt-radar: vorgehensweise zur identifikation von relevanten märkten und entwicklungspotenzialen emergenter technologien. In: Nyhuis P (ed) Wandlungsfähige produktionssysteme. Gito, Berlin, pp 419–456
33. Backhaus K, Erichson B, Plinke W, Weiber R (2006) Multivariate analysemethoden: eine anwendungsorientierte einführung (11., überarb. Aufl.). Retrieved from http://www.zentralblatt-math.org/zmath/en/search/?an=1114.62300. Springer, Berlin
34. Keeney RL, Raiffa H (1976) Decisions with multiple objectives: preferences and value tradeoffs. Retrieved from http://www.worldcat.org/oclc/2119417. Wiley, New York

# Chapter 6
# Platform Emergence in Double Unknown (Technology, Markets): Common Unknown Strategy

**Olga Kokshagina, Pascal Le Masson, Benoit Weil and Patrick Cogez**

**Abstract** The proposed chapter deals with platform emergence in double unknown situations when technology and markets are highly uncertain. The interest in technological platform development to enable creation of products and processes that support present and future development of multiple options is widely recognized by practitioners and academics. The existing literature considers that platforms already invented and the development is mostly based on exploiting this common platform core to build future markets and technological derivatives. However, when we are in double unknown situations, markets and technologies are highly uncertain and neither market options, nor platform cores are known. Thus, how to start an exploration? How can one ensure platform emergence in double unknown? What are the market and technology conditions that lead to different strategies of platform emergence? To answer these questions, we formally describe identified strategies and fabricate simple economical model to compare them. We illustrate the insights of the model through a case study of innovative technology development in semiconductor industry. Our results allow for better understanding market and technological conditions that allow for minimization of risks and exploration costs in double unknown and exploration costs in double unknown. Following the principle of value creation across various applications, this work extends the comprehension of generic technology design in double unknown.

O. Kokshagina (✉) · P. Le Masson · B. Weil
Mines ParisTech, CGS—Center for Scientific Management, 60 Boulevard Saint-Michel, 75272 Paris Cedex 06, France
e-mail: olga.kokshagina@mines-paristech.fr

O. Kokshagina · P. Cogez
STMicroelectronics, 850, rue Jean Monnet, 38926 Crolles Cedex, France

D. Cetindamar et al. (eds.), *Strategic Planning Decisions in the High Tech Industry*, DOI: 10.1007/978-1-4471-4887-6_6,

## 6.1 Introduction: Platform Emergence Strategies

Innovative technological development requires capital investment in its R&D. Yet, the R&D outcomes normally take years to be realized and the economic return is highly uncertain. In this uncertain environment project selection, choice of key technologies for the firm in upcoming years, management of R&D are of crucial importance for high-tech companies. Therefore, a lot of scholars and practitioners seek for tools to allow mitigating uncertainties and maximizing profit of future products in these breakthrough situations. The existing literature indicates the interest in developing technological platform to enable creation of products and processes that support present and future development of multiple modules [1–4]. This strategy appears to be robust for new product development, manufacturing, and distribution. According to [5], concepts of *modularity in products and organizations* are the core concepts driving the new kinds of product strategies emerging in dynamic product markets. Historical stories of IBM technological platform development [4], Black & Decker power tools [6], and many other examples show the relevance of platform strategies in practise. Meyer and Lehnerd [7] highlighted that the focus on single products or customers decreases commonalities in between products. At the same time, extreme market diversity and high level of nonplatform scale economy foster unified market, technology exploration [8].

Most often, in the literature platform (common technology) is considered to be known and the development is based on exploiting this common technology to build future modules. However, when we are in double unknown situations (unk unks, unforeseeable uncertainty [9]), markets and technologies are highly uncertain and neither options (modules), nor platform core are known. There is a possibility to decrease uncertainties by launching exploration trial projects for identified potential market needs and manage risks at the level of single projects. But the preference of high-tech industries like pharmaceuticals and semiconductors, is to develop common technologies relevant to several markets. This generic, radical technologies «yield benefits for a wide range of sectors of the economy and society» [10–12]. Generic technologies create potential for value creation across a broad range of industries.

On the one side, double unknown increases risks and it is hard to manage and evaluate projects. On the other side, double unknown situation allows to take into account several markets addressed by the same technology (platform), it gives a potential of working with several markets. Thus, how can one ensure platform emergence in double unknown?

Literature provides various definitions of platform. The essential part of platform is a set of common elements, especially the underlying core technology, implemented across a range of products [7, 13]. [14] states that the process of platform development starts with first, designing a common core. The platform core should account for market and option variety, low cost exploration, and adaptation [15, 16]. So a platform emergence begins with a common core design.

The first identified strategy of platform creation consists in progressively emerging the common core. Generally, this leads first to conduct exploration projects to create new knowledge and formulate platform, and then consequently reuse identified platform to construct platform derivatives (series of projects generate the technology platform with associated derivates) [17]. This strategy has to take into account risks of the chosen exploration project at the beginning, and consequently risks of portfolio options once it is constructed.

The second strategy creates a possibility to develop directly the technology that is more critical (generic technology) for several markets (platform derivatives) and not to seek for a concrete market at first. This strategy does not obtain common core as a result of exploration, but directly design the "common unknown" core as an object to conceive and manage.

So platform emergence can be organized by an exploration project emergence to fabricate platform or a direct platform emergence for identified options. In the first strategy, at the beginning, singular project is launched to find a solution to a particular problem and after to deduce common core based on a first trial exploration. This common core serves as a platform to build market applications. In the second strategy, we formulate the common core by identifying what is common to several options before trying any of them. But in the second case, the common is not necessary an existing one—it is common unknown to several potential markets. We will call the first strategy "Singular project Trial and Reuse" ($S_P$) and the second—"Common unknown" ($S_U$). Previous work described these strategies based on identified managerial framework with objects, actors, criteria, and necessary resources [18]. Still, even if highlighting major differences in between strategies, their economical conditions are not explicitly addressed. It is not obvious which strategy allows for better performance in particular situations. And moreover, the existing work does not show which metrics to use in double unknown to select a more suitable strategy (research gap).

The history of innovation promotes mostly singular project trial and reuse strategy to guide innovative development ($S_P$). But in certain sectors like semiconductors, we state that the development of generic technology (common core) is more important than singular project success. There are companies like Intel (PC Architectures), Microsoft (Operating system), etc., that were able to propose solutions for common technology before having developed any markets and then introducing products based on this common core to all the identified markets applications (common unknown development).

Thus, why common unknown strategy appears to be so challenging? Why do we usually prefer to conduct project exploration to emerge common core instead of constructing future common ground directly? What are the obstacles and advantages of each strategy? And which conditions one has to follow to choose one of them in organizing exploration activity? The objective of the chapter is to define what are the market and technological conditions that in certain situations lead to: (1) develop common building block (common core) that makes accessible all the other options but do not provide access directly to the market $S_U$ (2) launch singular project exploration to emerge future platform core $S_P$ (Research questions).

Generally, common unknown strategy $S_U$ has a lot of constraints and it is considered to have little chance to succeed. Among them there is capability to identify potential markets without adding too many constraints to the common core formulation, high cost of adaptation for future derivatives, competence to identify common unknown in high uncertainty. As all the markets are quite contrasted (different domains, criteria, etc.), it causes high costs of adaptation for each market module. The goal is to create a platform that reunites maximum functions in between different market modules. By considering various modules, we add more constraints in finding commonalities in between them, and probability of finding common core becomes almost zero. By adding specific market requirements, we increase the constraints and the risk of failure to find common unknown. Though even formulated theoretically, common unknown strategy appears to be risky, difficult to succeed and requires particular expertise in both technology and marketing.

Thus, is there common unknown strategy that allows platform emergence in double unknown and (1) avoids adding too many constraints by formulating common core (Hypothesis 1); (2) decreases costs of adaptation for market derivatives (Hypothesis 2)?

Besides the more classical $S_P$ strategy that leads to project exploration to formulate platform core appears to be risky as well. The strategy of project trial and reuse has a tendency to reduce the exploration space by formulating the context. This could result in a rigid platform core that is too specific for a chosen market.

We attempt to compare these two strategies by first, formally describing each one using simple analytical model. We will apply and evaluate the proposed framework on the case of innovative technology development at STMicroelectronics.

The chapter is organized as follows. First, we account for the existing risk management strategies to define strategies suitable for platform emergence management in highly uncertain situations. Second, we present chosen research methodology. Third, using proposed framework, we formally describe each strategy to be able to compare them and highlight their differences. We create a model that leads us to define the principles of technical and market compatibility to formulate the conditions of strategies applicability. We describe briefly the context of empirical study and we illustrate singular project trial and reuse $S_P$ and common unknown $S_U$ strategies using the chosen case study of advanced technology platform development. Finally, the chapter closes with managerial implications of platform emergence strategies and directions for further research.

## 6.2 Theoretical Background and Problem Formulation

We deal with management strategies in double unknown. To be able to characterize these strategies, we account for classical risk management techniques that deal with low uncertainty. Uncertainty is defined as a condition where decision maker does not know all the alternatives and there are risks associated with each alternatives or the consequence [19]. Under uncertainty, little is known related to the list of alternatives or the outcomes, the probabilities associated with the state of nature are unknown [20]. The economic success is influenced by many uncertain exogenous and endogenous factors. The future profitability depends as well on how the decision makers manage the projects after they have started and it is well demonstrated in the literature that continuous uncertainty management is required for innovative projects [21]. Literature review on risk management allowed identifying two types of strategies, which have a tendency to manage uncertainties at the level of projects or portfolios (platforms).

Sanchez et al. [22] showed that project risk management is a well-developed domain in comparison to the program risk management and portfolio risk management fields. They stated that for portfolio management it is hard to find particular written methodologies. In portfolios, usually we pilot risks case by case without considering an influence of project dependencies to overall portfolio performance. The risk management methods based on uncertainty reduction for identified projects are well represented (a lot of work deals with studies on how decision makers cope with uncertainties (i.e. [23]; projects with variations and foreseeable uncertainties in [9], etc.). Risk management includes techniques to increase either probability of occurrence of an event or its impact on the project (or decrease in case of negative risks) [24]. These strategies lead to minimize unknown by selecting a priori the less uncertain projects with higher probability of occurrence, depending on the identified market and technological risk. The level of uncertainty allows prioritizing corresponding markets (based on market probability) and selects a project associated with maximal economic performance (i.e., Expected NPV, Discounted Cash Flow). The risk management is concentrated on addressing uncertainties associated with project feasibility, market, technology, financial aspects, organizational implications, and so on [25]. At the project level, there is a tradition of uncertainty diagnosis and risk reduction for pre-defined problem (project context) ($S_1$). $S_1$ leads to minimize unknown by selecting a priori the less uncertain projects, depending on the identified market risks and technological risk. The criterion of "good" risk management is the high probability of success of the chosen project. These strategies deal with projects independently and do not consider the existence of common knowledge.

Risk management strategies in portfolio ($S_2$) try to take into account common aspects in between projects. The example of this is a portfolio represented by a technological platform core and its derivatives. This second family of strategies attempts to take advantages of interdependencies in between projects. For instance, in case of modularization, Baldwin and Clark [3, 16] propose to reuse the platform core that helps to address various options that are depending on it. Baldwin and Clark [16] showed how to obtain several options available thanks to common platform. Platforms represent a core of technological system and have to be interdependent with other parts of the system [4]. According to platforms typology [26], we deal with internal, inside firm platforms in this work. Reuse of platform core enables to minimize risks by constructing several options [16]. The module considered to be defined once the market signal is sufficient enough to conceive it

[2–4, 27, etc.]. This strategy ($S_2$) increases chances to succeed not by selecting one single most probable project but by increasing the size of the sample, i.e., by being able to play several options, maximizing the total economic value of the portfolio of derivatives. Risks are managed by a portfolio manager or a platform manager. The criterion of "good" risk management is the aggregated profitability of the portfolio (or platform), aggregated value or platform utility. This strategy works on common solution, but it is limited to the common already known aspects in between projects.

As was shown in [18], when it comes to highly uncertain situations (breakthrough, radical, disruptive, major innovation, etc.), $S_1$ might be impossible, because all projects are too risky. One still could make hypothesis (in case of unforeseeable uncertainty [9]) based on subjective probabilities of success, but they change significantly at the end of projects. $S_2$ strategy might be impossible, because there is no platform available to play several times with limited costs in high uncertainty. In addition, the existing literature on product platforms assumes that the platform leader knows the final use of products and is capable to develop these new products [26]. This is definitely not the case in the context of radical innovation when both the selection of platform core and final products are highly uncertain.

Nevertheless, a lot of researchers seek the way to deal with high uncertainty ("unk unks", unforeseeable uncertainty, etc.). Loch et al. [28] provide an overview of the existing strategies to unknown management. By showing that traditional risk planning techniques are insufficient for management of unforeseeable uncertainty, they suggested that the final method depends on the presence of unforeseeable uncertainty and complexity of the initial problem. Their work proposed a complementary model for diagnosis of unforeseeable uncertainty by learning problem structure and decomposing the problem. This research summarized two fundamental approaches for uncertainty management:

- Trial and learning approach [29–31] that consists of iterative trying selected trials and flexible changes in the course of action.
- Selectionism [29, 32, 33] consists of launching multiple trials in parallel and then selecting the best approach later. Selectionism is often considered to be more expensive and is affordable to use for big problems. Usually, selectionism is less time consuming than trial and learning and more suitable for market driven approaches that need faster response.

In the situation of high uncertainty and low complexity, they suggested to use trial and learning approach and in both high uncertainty and complexity—combination of trial and learning and selectionism [28]. In [18], it is shown that both these approaches lead to fabricate trial project to create knowledge that allows reducing uncertainty and identify alternatives. This leads to risk management first at the level of project and consequently at the level of portfolio once the trial project results in platform core. Hence, we acquire common core as a result of uncertain projects exploration that attempts to find a solution for a particular identified challenge. The developed common core serves as a basis of successive explorations. We call this strategy singular project trial and reuse ($S_p$). Empirically, Loch et al. [28] used combination of trial and learning and selectionism method of launching parallel trials in application to the case of Escend Technology start-up. Sanderson and Uzumeri [17] showed how generic platforms were able to coexist within the Walkman product family consequently (project exploration) and support the development of important subfamilies.

In [18], the second type of strategy that deals with double unknown situations was introduced. This strategy does not attempt to obtain common core as a result of exploration, but works directly with "common unknown" as an object to conceive and manage ($S_U$). The common unknown is precisely a common core for several identified market options. In common unknown strategy, there is a tendency to pay exploration phases that allows designing common technological core to enable access to several emerging markets exploration. Common unknown strategy to pilot risks was introduced based on the literature review and its application was illustrated on empirical case of technology development. This strategy does not deal with the same object as $S_P$, it uses different resources and requires particular conditions to its implementation. The actors responsible for its management have to take into account both technical challenges and emerging market needs and have a competence to link them in order to create common core. Even if identified based on the literature review, common unknown strategy is not widely used in practise since it is considered to be expensive, risky, and hard to manage due to several reasons. Among them:

- The capability to identify potential market derivatives in the case of double unknown.
- The capability to identify common unknown to several potential markets and construct common core.
- The cost of adaptation of common unknown to address particular market options has to be relatively small to ensure the profitability of the overall portfolio to justify preliminary investment for common core construction.

Thus, common unknown strategy creates particular condition for its implementation and in practice, the tendency is to select less risky development of exploration projects type $S_p$. Nevertheless, the more classical $S_p$ strategy that leads to project exploration to formulate platform core appears to be risky as well. First of all, high level of uncertainty in the phase of project selection (subjective probability of success that changes over time), can lead to develop project that will not be suitable for being platform core and address already identified markets. Studies of new industrial product failures consistently show that selection of inadequate market knowledge, brought about partially by ineffective market research, is a key contributor to the failure of project exploration [34]. Second, it is not certain that the knowledge created during the trial project exploration can be reused in the phase of derivatives development.

Therefore, it is not evident which strategy is more advantageous for innovative technology platform development in double unknown. The objective of the chapter is to precise what are the exact situations that make possible the identified strategies in double unknown, what are their limits and advantages.

We attempt to define what are the precise market and technological conditions that in certain situations lead to develop a common building block (common core) that facilitates all the other projects but do not provide access directly to the market. Hence, we seek the conditions in which common unknown strategy for platform emergence in double unknown:

**Hypothesis 1** Avoids adding too many constraints by formulating common core.

**Hypothesis 2** Decreases costs of adaptation for market derivatives.

To address our research questions, we attempt to define necessary criteria and resources to model each strategy, to characterize identified strategies and to investigate the applicability of risk management strategies in double unknown. An analytical model will allows us to understand the logic behind each strategy, its consistency and interest of each strategy in practice. We apply a proposed framework on an empirical case of advanced technological platform development in STMicroelectronics.

## 6.3 Research Method and Data

The objective is to analyze the precise market and technological conditions that in certain situations lead to platform emergence based on singular trial and reuse project exploration or common unknown identification. There are various ways to model risk management strategies behavior. The goal for our framework is not to explicitly model each strategy but to show in general the difference in between them, to show to which extend each strategy can be implemented in practise and to better understand major challenges and key success criteria of each strategy.

### *6.3.1 Analytical Framework*

To propose analytical model for $S_p$ or $S_U$, we start by characterizing the classical $S_1$ and $S_2$ strategies. Risk management strategies deal with both market and technology uncertainty [35]. Market uncertainty is based on the volatility of market size, customer needs. Technology uncertainty concerns the volatility of available knowledge and field set [36].

According to [37], product architecture can be described as the arrangement in between product's functional elements and physical components, including the specification of interfaces between interacting physical components. Various researchers used this approach to describe modular and integrated product architecture [38, etc.]. Regarding to this literature, each market product ($M_i$) can be described as a set of functional elements ($F_{ij}$) it targets and each technological solution is a set of physical components ($T_i$). To find a technical solution to a chosen market is precisely mapping in between chosen functional elements and physical components (Fig. 6.1). Hence, we could present each market $M_i$ as a set of key functions it targets ($F_{ij}$, where $j$—number of functions). Each function has a probability of occurrence $p_{ij}$ and a cost of its development $c_{ij}$ that depends on each market functions. Thus, each project is modeled as a technology development to address a set of desired functions for selected markets. Risk management strategies attempt to reduce uncertainties through this exploration and select a less uncertain, more advantageous project to develop.

#### 6.3.1.1 Risk Minimization at the Level of Projects—$S_1$

This strategy lead to minimize unknown by selecting a priori, the less uncertain dominant projects from the list of identified candidates $P_1,\ldots, P_i$, depending on the identified market risks. The level of uncertainty allows prioritizing corresponding markets (there is probability and potential benefits distribution for list of markets (see Fig. 6.1)): $M_1,\ldots, M_n$—list of predefined markets, $P_1,\ldots, P_n$—list of corresponding probability of success for each market accordingly: $P_1(M_1),\ldots, P_n(M_n)$. In $S_1$, marketing is able to prioritize market and predefine a dominant market to address with associated project definition containing information about desired functions, targeted clients, technical specification, potential profit, and so on. Cost of the projects exploration is $C_1,\ldots, C_n$. We consider that cost of each project exploration is high and with defined budget for a given time period one can pay only one project. So the cost of projects exploration is limited by predefined budget of R&D ($B$). So $C_1,\ldots, C_n \sim B$. The budget of a project is predefined (usually R&D budget for project development) and we consider that selected project $P_i$ cannot overcome it. $\Pi$ is the expected benefit. The expected benefits are determined on the basis of estimated a priori subjective probability function. The probabilities are

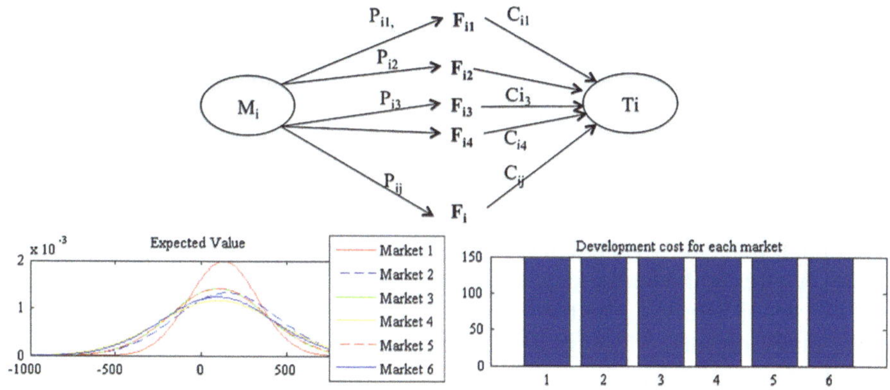


**Fig. 6.1** Principle of technology—market modelling

regarded as being subjective which helps to avoid restriction in business decision making [39]. The expected value is the weighted average and is found by multiplying each expected net present value (ENPV) by its respective probability.

The project manager normally selects a market candidate $i$ based on prioritized candidates list according to the ratio of the standard deviation and mean of expected value (coefficient of variation) or according to his utility function. The expected value of each project is calculated based on:

$$\text{ENPV}_{S_1} = \max P(M_i) \cdot \Pi(M_i) - C_i, \text{ where } C_i \leq B \tag{6.1}$$

The criterion of "good" risk management is the high probability of success of the project and the minimum coefficient of variation for project with controlled budget (or utility function depending on risk aversion). The coefficient of variation (standard deviation divided by expected value) is used to measure risks while comparing alternative projects. The higher coefficient of variation, the riskier is the project. The risk management is concentrated on addressing uncertainties associated with project feasibility, market, and so on. Risks in $S_1$ should be managed by project leaders that are capable to define and calculate information based on probability of success of different solutions, to reason based on both technical and market planning.

Once the project $i$ for market $i$ is selected the development is based on addressing the desired functions $F_{ij}$ by identified technical solutions $T_j$ while keeping costs under control.

#### 6.3.1.2 Risk minimization at the level of portfolio—$S_2$

As was described before, risk minimization at the level of portfolio consists in using an existing platform core to construct several options ($S_2$). This strategy increases chances to succeed not by selecting one single, most probable project, but by increasing the size of the sample, i.e., by being able to play several options, maximizing the total economic value of the portfolio of derivatives. Risks are managed by portfolio or platform manager. The criterion of "good" risk management is the aggregated profitability of the portfolio (or platform).

In platform driven strategy, platform core is considered to be predefined—$P_{\text{of}}$. In [16], the total economic value of the system is expressed as the sum of a minimal given value (platform core in $S_2$) plus the incremental value added by each module. The objective is to construct market derivatives that are based on $P_{\text{of}}$. Strategy responsible has to select a platform core from the list of given already (identified) potentially candidates. In the case of $S_2$, we can have several potential platform cores and take the one, which has a higher expected value.

Once $P_{\text{of}}$ is selected, one tries to reveal potential market candidates that could reuse platform core. Usually, considered markets have equal rather than low probability, so one can play several options. Considered probabilities of market derivatives are normally low, and therefore these markets are not advantageous for

$S_1$ strategy being $P_1, \ldots, P_n < 1$ (low probability equal for different derivatives, where $n$—number of project derivatives). Aggregated cost value of market derivatives development has to be slightly low and reuse the already existing platform core. Each option attempts to address different market derivatives maximizing the total economic value of the portfolio itself. The constraint is defined by budget: $\Sigma C_i \sim B$.

Expected system value is defined as an expected aggregated portfolio (or platform) profitability:

$$\mathrm{ENPV}_{S_2} = \sum_i \mathrm{NPV}(P_i) = \sum_i (P(M_i) \cdot \Pi(M_i) - C_i) = n \cdot p \cdot \Pi - B \quad (6.2)$$

We suggest that in the case of low probability of $n$ market derivatives, we can consider them equal: $\sim p$. Portfolio manager has to know well the platform to identify derivatives. He has to manage the portfolio of options and verify that the sum of options is profitable.

As was stated before in the situation of radical innovations, one cannot predefine dominant market like in $S_1$ and it is impossible to select a platform like in $S_2$ to address platform derivatives. The literature leads to identify the singular project $S_P$ strategy that starts as project driven and then attempts to reuse identified projects as a platform core.

#### 6.3.1.3 $S_P$: Singular Project Trial and Reuse Strategy

Traditional risk management criteria like (Discounted Cash Flow, Internal Rate of Return, Cost-Benefit Ratio, etc.) assume the predefined value of uncertainty over the course of projects development, and thus do not consider flexibility necessary for projects in double unknown. In the situation of radical innovation, we cannot select a market with higher probability and implement $S_1$ strategy. As well, we do not have predefined platform core to address potentially emerging markets. Thus, we are dealing with double unknown when markets and technologies are unknown.

Markets are considered being unknown since we cannot predict their volume and value and probability distributions are volatile. However, we are still capable to predefine main functions ($F_{ij}$) that future market ($M_i$) seeks to address. According to consumer behavior theory [40], "goods in combination may possesses characteristics different from those pertaining to the goods separately" and therefore can create new needs. The emergence of new markets can be modeled as a combination of various functions. For example, taking an energy harvesting, that is a common interest for a lot of different sectors, we could already announce energy control, transformation, or energy storage as possible functions. The same for the technology: even if we do not know which architecture, form, size, performance, and so on it will have exactly, we could list potential candidates for future functions. Hence, in double unknown, it is possible to work on the set of functional combination for technologies and markets and the exploration in double unknown can be seen as a matching in between potential functions and physical elements that can address them (emerging technologies).

As the payoffs (future benefits) and costs are unpredictable, one will have chance node that reflects several probabilistic payoffs and cost scenarios. This implies that the costs distributions ($C_i$) and future benefits will have a non-zero variance. To reflect managerial flexibility, one could embody decision nodes that enable management to eliminate poor alternatives to avoid bad outcomes. The uncertainty affects the expected investments values by widening the distributions [41]. Thus, the market probability distribution in the high uncertainty situations are considered almost flat and in the first phase of $S_P$ is hard to prioritize markets (as in $S_1$). We consider that it is still possible to prescribe subjective probabilities in order not to select dominant market, but estimate which of them can be less risky and more accessible for future market derivates. Therefore, we select less risky project (trial project) based on the coefficient of variation of the expected project value:

$$\mathrm{ENPV}_{S_p} = \max P(M_i) \cdot \Pi(M_i) - C_i, \text{ where } P(M_i)_{S_P} \leq P(M_i)_{S_1} \tag{6.3}$$

It is important to mention that at the beginning of $S_p$, we are not looking for common core to construct platform but more to address uncertainties relevant to chosen market. The project exploration results in platform core construction. We complete this strategy by the second phase of platform reuse to develop identified market derivatives: $M'_1,\ldots, M'_n$. Project $P_1$ developed in the first phase is becoming a platform core $P_1 = P_{\mathrm{OPF}}$ that makes accessible derivative markets in the second phase. The knowledge created through the first phase attempts to decrease uncertainty relevant to other projects. The total expected value of this strategy is the sum of the expected value of the first exploration project and aggregated expected value of project derivatives.

$$\mathrm{ENPV}_{S_{p_1}} = \sum_i (P(M_i) \cdot \Pi(M_i) - C_i) \tag{6.4}$$

where aggregated value contains the value of both exploration project and market derivatives.

The risk management in this combined strategy depends mostly on trial project identification and exploration in the first phase. Platform core selection is limited to the first developed project context and its flexibility for identified future options. In $S_P$ we assume several risks: we accumulate uncertainty relative to trial project selection that can result in platform core and risks associated to derivatives management. For project manager in $S_P$ there are the same risks as in $S_1$ and $S_2$, but it is much more challenging to identify exploration project in high uncertainty. In addition, the budget required for the second phase strongly depends on how well the trial project was identified and managed.

#### 6.3.1.4 $S_U$: Common Unknown

The major difference of $S_U$ in comparison with $S_P$ strategy is that we attempt to design solution directly as a common unknown for potentially emerging markets. Instead of starting with project exploration to reuse maximum known, we start directly with platform identification as common unknown to several market derivatives. As we saw in previous research there are situations (in emerging fields), when it is possible to identify dependencies in between emerging future markets and construct common core in between them. These dependencies are based on market and technology to identify common unknown. We characterize market interdependencies as common functional characteristics for set of various markets. The logic of common unknown is to fabricate common core for the list of potential market candidates with low probability. Then common unknown is a technological building block to address certain specific functions common to the set of identified markets.

The risk management criteria based on uncertainty reduction (max value with min deviation) are not explicit for common unknown strategy. We are dealing with common unknown, where it is impossible to highlight probabilities for markets and technologies that do not exist yet. Instead, we are increasing the variety of options to play. We construct common core based on potential commonalities in between emerging markets. In this strategy, we launch a preliminary phase of common unknown identification that will address all the identified options. There is a preliminary cost of paying $P_0$ project that enables to significantly reduce the cost of each option development.

In double unknown, it is necessary not just to minimize uncertainties for selected exploration space, but also to maximize the variability of future options to play in order to have a generic technological core. There is certainly risk associated to common unknown identification. The total expected value of the platform developed:

$$\mathrm{ENPV}_{S_C} = \sum_i (P(M_i) \cdot \Pi(M_i) - C_i), \text{ where } \sum_i C_i + C_{P_0} \leq B \tag{6.5}$$

The aggregated cost of development of all the alternatives and the cost of common unknown exploration should be to the budget? article of R&D.

### 6.3.2 Comparative Case Study and Data Collection

To explore the proposed model utility, we carried out a field study in semiconductor silicon foundry and analyzed an empirical case of technological platform development to compare identified strategies. The relevance of semiconductor industry for radical innovation studies was showed by various researchers (e.g., [42]), especially for knowledge creation methods in science-based environments (showed by [43, 44]) driven by "More Moore" Law [45]. Strong competition, fast changing environment relevant to semiconductor industry lead it to explore not just new technologies, but as well new functionalities, creating new products. Semiconductor industry normally accounts for generic technologies being designed for various market applications.

We conduct our case study at STMicroelectronics, one of the leading semiconductor companies, in Advanced R&D research units that do not follow classical rules of R&D management. Advanced R&D units in STM do not follow "More Moore" law. They are subscribed in diversification approach that is identified by ITRS as "More than Moore" [46]. There is neither clear scientific question, nor well-defined decision to develop new products based on exploration and targeted markets. There is high level of uncertainty both at the level of technology and future markets.

To better understand the applicability of $S_p$ and $S_U$ strategies in double unknown platforms, we illustrate their application on the analysis of advanced technology platform development. We adopted a longitudinal case study design [47]. This case of technology development implied portfolio of Collaborative R&D projects, PhD thesis, and business unit development projects. For this empirical study, the primary source of data was regular and frequent semi-structured interviews. We organized interviews with specialists participating or leading technology development from R&D technology and design groups, business divisions, former PhD students, and some associated external research centers. Overall, around 30 interviews were performed. The analysis was completed by the scope of documents as European projects reports, research presentations, thesis manuscripts and database of thesis project descriptions. In addition, data analysis was followed by seminars with company managers (not necessary participating in technology development) to discuss the project, to test the validity of our hypothesis, and enrich our propositions. This work was conducted over 8 months period from (November 2010 to June 2011).

The studied case is BICMOSMW (high performance 0.13 μm SiGe BiCMOSMW technology, targeting very high-frequency applications). It is a technology platform development based on heterojunction bipolar transistor (HBT) with unique technology features. Despite of the difficulties in defining both future technology and designing market, the team succeeded to address several markets simultaneously.

The choice of a case study was argued based on several criteria: both market and technological unknowns at the early stage of technology concept (double unknown situation), the generic aspects of technology functions, potential of technology to address several market applications (technological platform development possibility), and the data access for the whole set of ex-post project analysis.

In addition, the case is used to analyze the differences in $S_p$ and $S_U$ due to the some redundancies in the management approach identified through the conducted interviews. Based on the interviews conducted we have found out two versions of technology platform development that correspond to previously presented $S_p$ and $S_U$ strategies.

On the one side, first group of interviewers was presenting the case as management of exploration trial project of BICMOSMW technological development for radar application development (automotive applications) and its reuse for fast download (wireless communication) and optical communication applications in the following steps. On the other side, the experts who were responsible for the technology design and management showed that it was not executed completely neither at the level of project based strategy, nor at the level of portfolio-based exploration. The team leader concentrated the platform exploration in addressing the commonalities for identified emerging markets applications. He was seeking precisely for common unknown. The promising interdependencies were constructed based on common unknown identification ($S_U$) and were managed based on the links to allow exploration. These different opinions in the case representation lead us to apply our analytical model using both strategies to BICMOSMW case of technology development to better examine them.

Further, we illustrate singular challenge $S_p$ and common unknown $S_U$ strategies using empirical material to highlight the major differences, specifics, and prerequisites to launch these strategies.

## 6.4 Results

### *6.4.1 Case Description BiCMOSMW Platform Development*

We illustrate the identified risk management strategies based on BiCMOSMW (high performance 0.13 μm SiGe BiCMOS technology, targeting very high-frequency applications) technology platform development [48]. Chantre et al. [49] showed that high-speed BiCMOS roadmap is driven, on the one hand by the increase of the optical communications data rate, and on the other hand by the emergence of applications at higher frequencies. Thus, it does not follow classical More Moore Law.

In STMicroelectronics, BICMOSMW platform has evolved after several generations of technical solutions. The history of bipolar transistor technology based on SiGe at STMicroelectronics started in 1998 with 0.35 μm technology for wireless communication [50]. The success of the SiGe HBT has come from its compatibility with silicon technology allowing both low cost and high yield.

Till 2002, the group was working on optimization of bipolar transistor for analog signal processing to address emerging standard of 60 GHz in order to enable wireless communication. There was no particular client demand at the beginning. The developed technological basis was not ready to address any market at that time. In the following 2002, the group was analyzing which potential high-volume market and technological efforts are needed to develop while reusing the previous research results on bipolar transistor.

The expert (Technology Line Manager) that initiated technology development was looking for an emerging market with potentially huge volume to ensure high return on investment. He identified a particular system issue: the Wi-Fi connectivity for big public systems like airports, train stations, and more generally for high-density areas with a lot of connectivity devices. The current issue considered was with the existing standard for Wi-Fi communication (2.5–5 GHz). The frequency of processing information allowed by standard was too low to ensure connectivity for each device and enable information security. One of the potential solutions was to use a 60 GHz Wi-Fi system with a long range (>10 m) to limit the number of base stations and system complexity. However, this standard was just emerging at that time.

Developed technological platform addresses various functions as high-frequency emissions, low-power consumption, digital signal treatment, covered distance, and so on. In the following, the technology platform developed (BiCMOSMW) did not allow the creation of initially identified "Wi-Fi for airport" system. Nevertheless, it served several market applications as automotive radar, optical communications, wireless fast download systems, high-speed instrumentation, and non-invasive imaging. This corresponds to common unknown strategy for platform emergence.

The difference in management aspects of the other part of the interviewees consisted in selecting market trial and reuse strategy. For them the first market application considered was automotive radar. The technology platform was specifically designed for this market and further reused for other applications. Hence, from one side it seems project was managed by $S_U$ type of strategy or from the other side by $S_P$.

Further more, we demonstrate the specifics of case interpretation by trial and reuse and common unknown strategy using our analytical model.

## 6.4.2 Data Analysis

We consider that each market follows normal probability distribution (evaluation of mean and standard value) to show the evolution of market probability distribution in the course of technology development. Due to the high uncertainty level all the market distribution curves are flat. To simplify the calculation, we use discrete probabilities $P_i$ and we suppose that all the markets have the same volume $V$. Initially, six potential markets were identified with associated probability distribution for BICMOSMW. As uncertain emerging markets in this chapter, we consider markets with new added functionalities (new products) or completely new markets: $M_1$: radar (automotive market); $M_2$: optical communication; $M_3$: standard linear products; $M_4$: medical; $M_5$: hard disk drive applications; $M_6$: fast download.

We suppose development cost for technology for each market is equal to $C_i$ ($C_i = 150$). In this work, the development cost is the cost of technological building block only. We consider that the design, test, package, interconnections, mode of implementation, etc., are always necessary for the final product development and their costs do not vary significantly with the features of technical core in semiconductor industry. So we do not take these costs into account (they are inevitable for product development in each sector). The defined budget of R&D for technology platform development is considered $B = 250$.

#### 6.4.2.1 Singular Project Trial and Reuse Strategy *Sp*

As we are in the situation of high uncertainty, all the markets have really low probability and high volatility (Fig. 6.2). The markets probability distributions are almost identical. Therefore, it is hard to select one most promising project. Nevertheless, market analysis showed that the radar application was more certain and the team started to explore potential bipolar technology targeting automotive market. By decomposing the projects, they defined functional requirements ($F_{1j}$) relatively to the automotive market: low consumption, high frequency, cover long distance, and integrated solution (CMOS + bipolar) (Fig. 6.3). The cost to develop the first project was determined as $C_1 = 150$. The technology was developed to address these functions. According to the strategy, once technology is developed, we seek if there are other markets that can potentially reuse it as a core to develop new products.

This gives us two scenarios:

**Scenario 1** Technology developed in the first phase of $S_p$ allows decreasing uncertainty relatively to other projects.

For example, high frequency of operation ($F_{13}$) and low consumption of technology ($F_{11}$) developed for automotive radar application, are relevant to optical communication and fast download system as well. So the exploration project $P_1$ decreases uncertainty (in Fig. 6.4 expected value distribution for projects 2 and 6 are less volatile) and development costs (the bars in red for the cost of

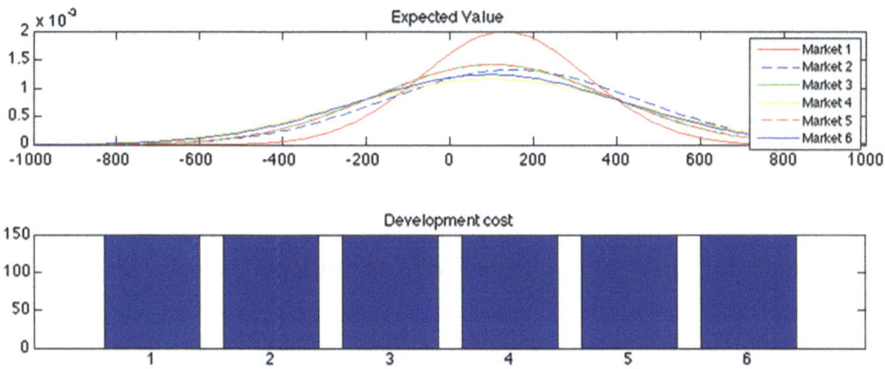


**Fig. 6.2** Initial situation for identified markets (*expected value* distribution and *development cost* for identified projects)

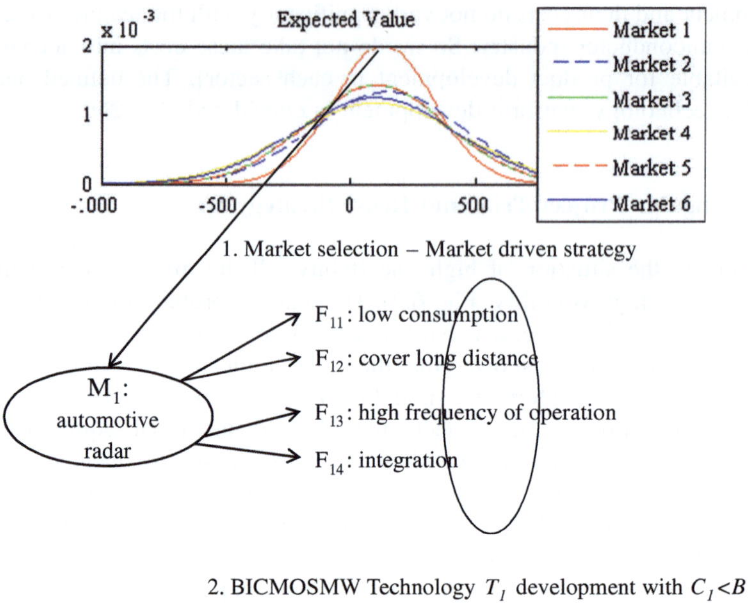


**Fig. 6.3** Trial project selection *Sp* strategy

projects 2, 6 are subsequently reduced thanks to $P_1$ reuse) relevant to the $P_2$ and $P_6$ projects. The challenge is that the aggregated cost of development for identified derivatives $P_2$ and $P_6$ should be inferior to $\sum C_i \leq B - C_{P_0}$. With the $B = 250$ and $C_1 = 150$. The $C_2 + C_3 = 100$. For the budget that rests, one can develop identified market derivatives (Fig. 6.4). The rest of development addresses remaining market functions. For optical communication ($P_2$) solution there was a need to ensure high voltage capability for optical communication and low noise, as well as high-performance passives in addition to functions reuse from $P_1$. Hence, the team had to redesign the system developed in trial project to adapt it to the new market with the cost of exploration $C_2$. For fast download ($P_6$) one had to ensure emerging 60 GHz standard with assigned cost of adaptation—$C_6$ (Fig. 6.5).

**Scenario 2** Technology developed in the first phase of $S_p$ did not allow decreasing uncertainty for other potential market derivatives.

This is the extreme scenario when technology explored for radar communication was specifically oriented to automotive industry. Therefore, the level of integration of developed technology and knowledge created during project exploration were not relevant to any of other market options identified, the developed core is rigid.

In this case, the level of uncertainty relevant to other options after radar application development does not change, and therefore the cost of development remains the same. The project fails to form a platform core since the exploration space was restricted by problem setting in the first place. Project derivatives cannot

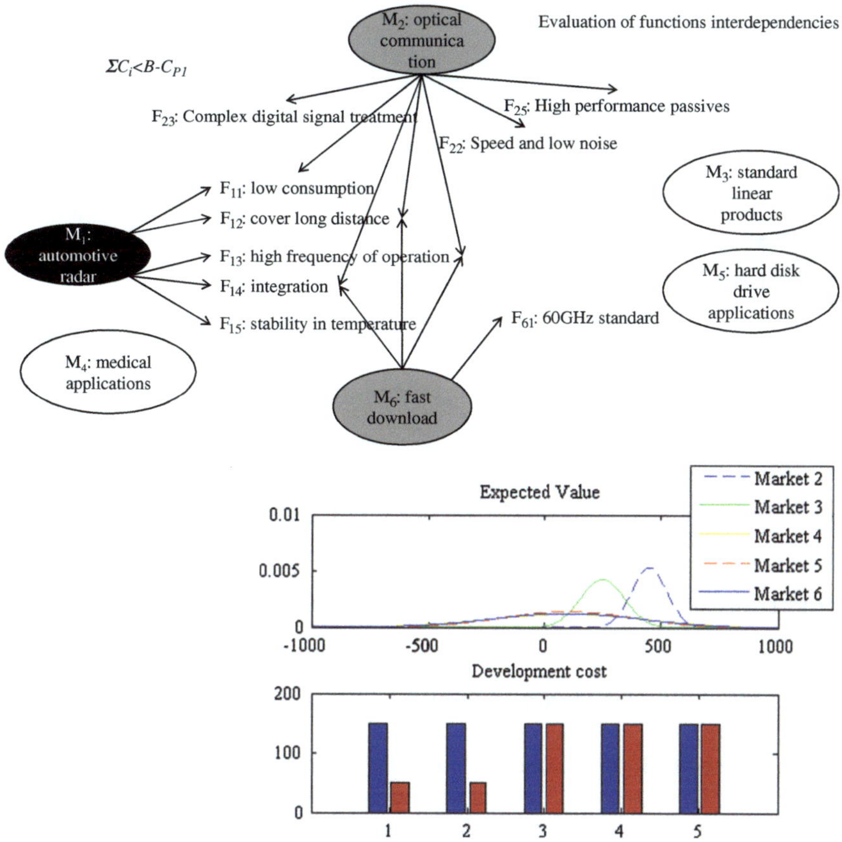


**Fig. 6.4** Market modules identification

reuse knowledge created to decrease uncertainty and the cost of adaptation is still high. Even if there is a failure in platform development, this scenario can still generate benefits if the radar application generates profits by itself.

### 6.4.2.2 Common Unknown $S_U$

We consider the same identified market distributions and the same costs at the beginning as in $S_p$. But there is an expert that proposes a project of technological development common to all identified markets $P_0$. He seeks for desired functions for all identified market derivatives and look for a technology to address them and be used by the set of market applications. Thus, future common basis to six emerging market applications to enable connectivity is precisely common unknown (Fig. 6.6). If it is possible to identify and develop $P_0$, we arrive to decrease the adaptation and further development costs sufficiently. In this case,

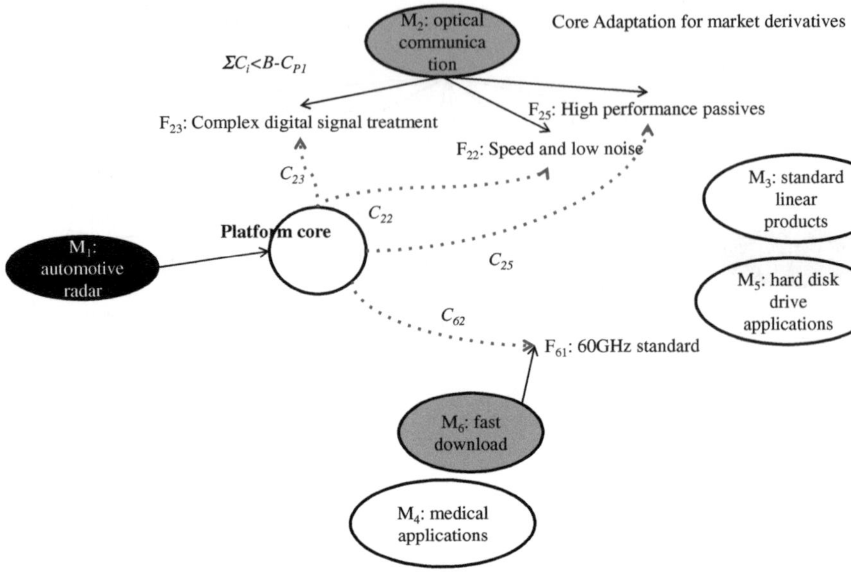


**Fig. 6.5** Technology platform developed followed by $S_p$ strategy

there is a possibility that there is a common core which is not known a priori, but its creation provokes significant minimization of the cost for other projects—the common unknown. In this example, common unknown $P_0$ is precisely "Connectivity for high density areas". The technological phenomena developed addresses maximum functions with the particular budget associated to access all the defined markets. The developed technology BICMOSMW allowed to explore maximum functions with fixed budget of R&D and addresses several markets at the same time. *Common unknown* was identified as knowledge gap that would connect knowledge related to functions of the future system that enables high connectivity (Fig. 6.7). It is important to mention that $P_0$ project does not target any markets but their commonalities. In Fig. 6.7, common core targets dependent functions as integration.

$F_{14}$, low consumption $F_{11}$, high frequency $F_{13}$, and long distance $F_{12}$. These functions are common for markets $M_1$, $M_2$, $M_3$, $M_4$, $M_6$, and $M_8$. Thus, for common unknown strategy to work one needs to identify common core and invest in one technology that will sufficiently decrease the cost of development for all identified markets (Fig. 6.7 bars in red). Suppose that $B = 250$. If the cost of development for $P_0$ project: $C_1 = 190$, then the budget left could be used to address market derivatives: $B - C_{P0} \geq \Sigma C_{Pi}$. By developing the common core that was precisely BICMOSMW for WIFI airport system, they have decreased the cost of adaptation sufficiently and enable the development of other options with the rest of the budget. This strategy leads to a more generic technology development than in the first case. We can take into account more market options to design core.

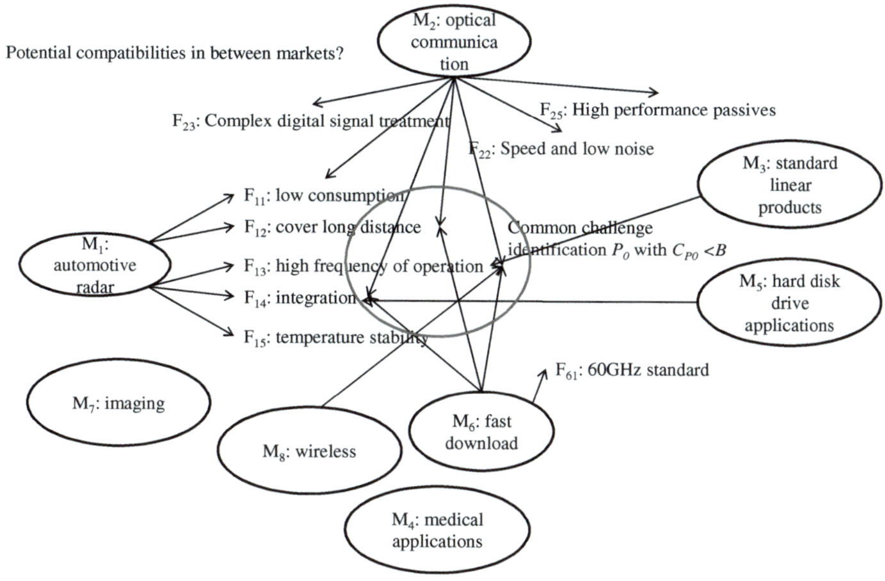


**Fig. 6.6** Market functional compatibilities identification $S_U$

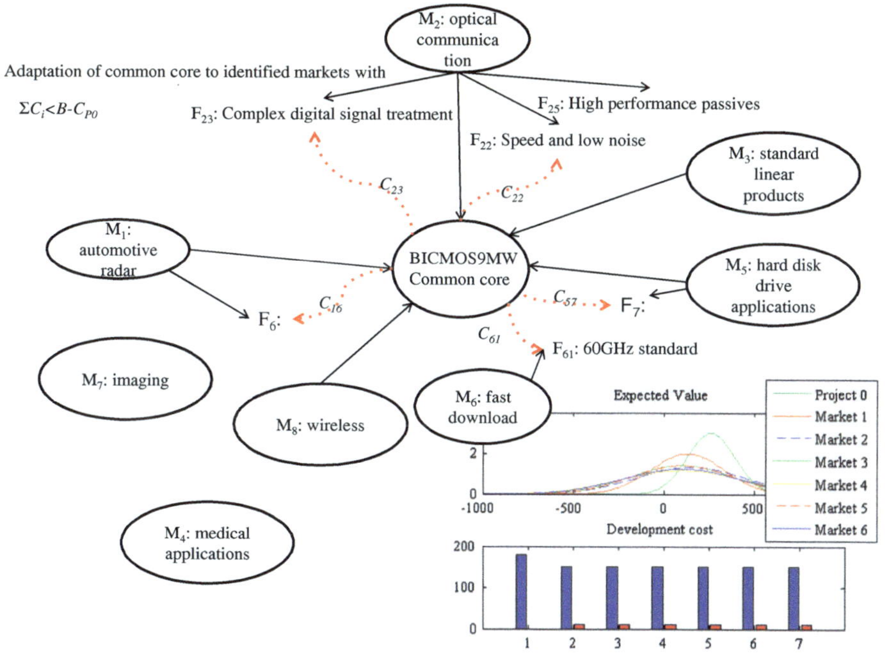


**Fig. 6.7** Technology platform development followed common unknown strategy $S_U$

This analysis shows that both strategies can result in platform creation. However, the example demonstrates the important differences in terms of risks and performance of both strategies.

## 6.5 Theoretical Propositions and Findings: Laws of Market and Technology Compatibility

Based on the proposed simplified framework, we see that the success of both strategies depend on the first project identification in $S_P$ and common unknown to design $S_U$ identification $P_0$ in the second strategy. Through this study we revealed that:

- It is necessary to compare the probability of success of technical development of common unknown $S_U$ to the probability of the first exploration project in $S_P$. The success of the strategies depends on the probability of $P_1$ (in $S_p$) and $P_0$ in $S_U$.
- The success of common unknown strategy depends on the effect of project $P_0$ to decrease costs of market derivatives projects $P_1,\ldots, P_6$. While launching common unknown strategy, we do not have to look for $P_0$ with minimum cost of development, but rather $P_0$ that allows developing all the modules with the cost inferior to the budget. Hence, we need to find common unknown which ensures the aggregated cost of modules development and common unknown exploration inferior to predefined budget of R&D.

These findings allow us to come back to initially formulated hypothesis for common unknown strategy properties. Common unknown strategy which allows platform emergence in double unknown should (1) avoid adding too many constraints by formulating common core (Hypothesis 1); (2) decrease costs of adaptation for market derivatives (Hypothesis 2).

**Hypothesis 1** Common unknown core avoids adding too many constraints in formulating core.

Normally, the more market alternatives are considered, the less is probability of commonalities identification in between them. It means that the probability to address simultaneously N markets $\sim 1/N$. To construct common core, we attempt to predefine future links in between different market functions. For example, for 2 markets we can have at least 1 link in between two functions $F_{ij}$, for 4 markets at least 6 links to ensure full connectivity. We suppose the connections in between markets can be developed with likelihood costs. If we consider new market for common unknown strategy, we need to validate if it is compatible with all the other markets. Number of compatibilities to ensure in between different markets: for 2 markets—1 link of compatibility, for 3 markets—3, for 4–6 links, etc. For $k$ markets, the probability of market $k$ to be compatible with the others is equal to $P^{\frac{k(k+1)}{2}}$. Thus, normally the more market alternatives we have, the less is probability of common core existence. In this case, it is extremely challenging to find common core for several potential markets due to very low probability of the markets existence.

Through our case study, we have seen that the success of common unknown strategy does not depend on the number of identified markets. In the case of common unknown the probability of common core existence $p = P_0^* = \text{const}\ \forall k$. For instance, the probability that we find unique BISMOSMW for 6 markets or 2 markets is the same and equal to $P_0^*$ which is almost zero for each market in double unknown (due to high level of uncertainty). We formulate the law of markets compatibility $P_0 = P_0^* \cdot p^{\frac{k(k+1)}{2}}$ where $P_0^*$ probability of common core existence and $p$—probability of all the links to ensure for $k$ markets.

For example, taking six market applications identified for BISMOSMW technology, we suppose that the probability of each market existence is inferior to 20 % in case of double unknown. In that case, usually one waits for uncertainty decrease and launch development once the probability is higher. In the presented industrial case, they were not waiting for the particular market emergence but for the emergence of at least one market at the end: $1 - (1 - 0.2)^6 = 1 - 0.262 = 0.738$ which is more than 73 %.

Therefore, for uncertain markets with low probability of success one needs to ensure high level of compatibility in between functions, $p \sim 1$. In Fig. 6.8. High level of compatibility is ensured by a common core that uses four identified functions to construct a common $P_0$ core. In semiconductors, it is not a rare case since technologies have a general character and often one function could be reused in between different products (e.g., technology that targets high frequency is used by all identified markets in our case). In this case $p = P_0^* = \text{const}\ \forall k$.

We show that the common unknown is possible when compatibility of common core is high for all the identified markets. It means the probability of being capable

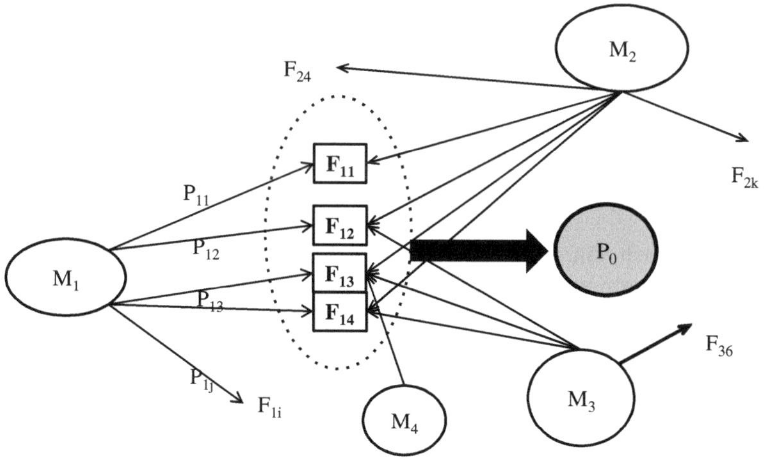


**Fig. 6.8** Markets compatibility

to develop the interdependencies or account for the general functions is high. We formulate the law of markets compatibility to construct common unknown:

*The law of market compatibility*: the probability of common core existence doesn't depend on the number of market alternatives; but on high probability of compatibility in between markets. $P_0 = P_0^* \cdot p^{\frac{k(k+1)}{2}}$, when p ~ 1 $p = P_0^* =$ const $\forall k$ for common unknown (Fig. 6.8).

**Hypothesis 2** Common unknown core that keeps cost of re-adaptation low for evolving market modules.

The second challenge for technology development in high uncertainty is to keep the development cost under budget constraints. In the $S_p$ strategy, the project development cost is first of all trial project exploration cost. When we develop trial project there is a possibility to decrease the cost of development for other projects. Thus, in the second phase of $S_p$ a development cost (due to minimized uncertainties and knowledge reuse) can be smaller for market modules or do not change depending on the results of first phase exploration. In the case of BICMOSMW, the trial $P_1$ project launched in first phase decreases the cost of the second and third projects. However, there is a risk that the trial project results in a too specific technology that is impossible to reuse.

While starting technology development, we do not usually look for $P_0$ project of common core development that does not address any particular market. The cost of $C_{P0}$ also could be superior to the cost of $P_1$. However, once we find $P_0$ project that allows decreasing uncertainties common to other projects ($M_1$, $M_2$, $M_3$, $M_4$ in Fig. 6.9), we are much closer to identified markets. The development costs for all the dependent options are much smaller (the cost of each market development $C_{Mi}$ could be decreased till adaptation cost ε). In common unknown strategy, there is a need to pay preliminary cost $C_{P0}$ for common core construction, it allows obtaining the cost of adaptation ε that is significantly smaller than the cost of development of each of these alternatives $C_i$: $\sum (C_{P_1} + \cdots + C_{P_N}) \geq C_{P_0} + N \cdot \varepsilon$.

Thus, we formulate the law of technology compatibility: one can pay technological common core development if it allows addressing all the identified options with the cost inferior to budget of R&D $\sum (C_{P_1} + \cdots + C_{P_N}) \geq C_{P_0} + N \cdot \varepsilon$.

How can we ensure costs of technology adaptation to be low? We can guarantee costs to be low if there is a high level of compatibility in between markets and we attempt to share knowledge among them. Launching development project by project, we do not seek for connections in between these projects. This could result in specific for each market technology development that is hardly reusable for others. While developing a technology for one application, the purpose is to develop a solution that generates profits right away and not the one that has potential to connect several ones. Certainly, at the level of mature project development and established markets, one would not look for technology and market compatibilities but for dominant application that generates benefits right away. However, in the early phases of technology investigation in double unknown, one could identify potential market compatibilities with related technological

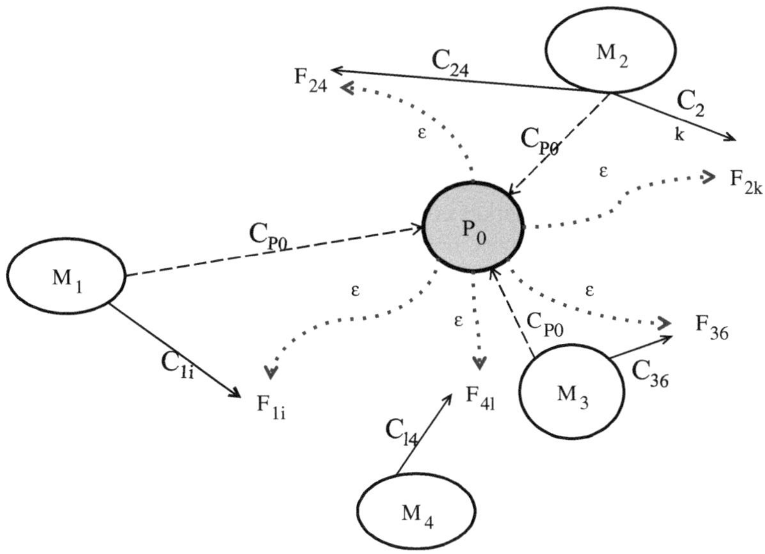


**Fig. 6.9** Technology compatibility

functions. Common unknown strategy could be advantageous when there are at least two interdependent potentially market solutions that are based on the common core reuse. For example in the automobile industry, the number of alternatives is restricted to particular geographical regions, different economies and most probably common unknown strategy will not be suitable. On the contrary, the enormous amount of technological alternatives and modularity in semiconductor industry makes common unknown strategy advantageous.

## 6.6 Discussion and Contribution

This chapter contributes to the literature on technological platform emergence in double unknown situation when technology and markets are highly uncertain and neither platform core, nor options are known. The objective of this work was to provide insights to a question of platform core emergence in double unknown. It is critical for strategic decision making to be able to choose whether to specialize on a concrete market—technology couple, or launch a platform development [3, 51]. Being able to make right decisions could help to avoid expensive worthless explorations [34, 52]. Through two identified strategies of platform emergence, this work advances the understanding of market and technological conditions that allow selecting a strategy and organize exploration in double unknown.

We have compared two possible strategies of platform emergence in double unknown (1) consequent emergence of project relatively to singular market challenge followed by a platform core creation that address market derivatives (Singular project trial and reuse) and (2) platform design based on common core identification relevant to several markets (Common unknown).

To investigate the question of market and technology conditions of each strategy implementation, we formulate a model that synthesizes the description of each risk management strategy for platform emergence. Thanks to this model, we introduced technology and market compatibility rules that highlight the situations when common unknown strategy is possible.

We characterized what are prerequisites that in certain situations allowed developing technology core to enable platform creation but do not access directly to the market. We introduced two laws of technology and market compatibility to enable the existence of common unknown strategy.

- *Law of market compatibility*: probability of common unknown core existence does not depend on the number of market alternatives, but on the high probability of compatibility in between markets.
- *Law of technology compatibility*: one pays the technological common core development if it allows addressing all the identified options with the aggregated cost inferior to budget of R&D.

*While guaranteeing* these two conditions, one can rely on common unknown strategy to exist when all the markets have equally flat probability of distribution in double unknown.

If one of the markets is more promising in terms if benefits and there are dependent markets that potentially can reuse the technology developed to this singular market, the singular project trial and reuse strategy is more attractive. It is certainly more advantageous for explorations when dominant application is identified. There is a negative effect that the results of project exploration can fail to decrease uncertainties relevant to other options. The study demonstrates that the strategy of singular project trial and reuse development for platform core emergence highly depends on the context of the first project setting.

As it has been shown, the common unknown strategy is relevant to the industries with various alternatives, mostly for upstream type of industries like semiconductor foundries. Following the principle of value creation across various applications, this strategy gives an opportunity to conceive generic technology exploration relevant for several markets in double unknown. The technology is designed based on functional compatibility and future potential of technology to decrease a development cost.

Even if some academics and practitioners insist on independency of advanced research exploration from addressing concrete markets issues, our work suggests the managerial implication of strengthening the collaboration in between technology R&D and markets divisions. This collaboration will allow the construction of flexible commonalities and enhance the value of common core relative to future market derivatives. However, the common core emergence has to be done in the earlier phases of technology exploration and market emergence (breakthrough innovations) to avoid too many constraints added by each market derivatives, because this will lead to rigid core construction and increase the cost of development. Common unknown strategy development requires high level of management expertize to organize and conduct exploration. Manager of common challenge should have a competence to identify common challenge and more precisely an ability to detect functional compatibility and capacity to reason based on future potential of technological solution. This requires proficient marketing skills to list all the potential functions and technical knowledge to address them, as well as a clear strategic vision.

Both presented strategies of risk management in platform emergence allow organizing a collaborative platform design. Using originally, new way of risk management based on knowledge gap identification to construct common core (Common unknown strategy) company can build its innovative capabilities through knowledge management and better position to innovate in emerging fields. Common unknown strategy maximizes the value exploration space while keeping costs under budget control.

The existing literature reveals that platform design is not about finding a common agreement but to create a common interest [14, 53]. Studied in these work strategies of platform core emergence allow to create this common interest or through trial project exploration or through common concept exploration in double unknown.

Finally, the conceived platforms that take into account several emerging market modules are more probable to avoid obstacles as changing boundaries of exploration, co-specialization as a complementarily versus mobility, etc. [54]. Flexible common core resulting after common unknown strategy development is more suitable for scope change thanks to its generic character.

The proposed model is static and its objective is to understand major differences in between strategies behavior regarding development costs and market probabilities. Further research will need to sophisticate our model to take into account specifics of real situations (competition, stochastic evaluation, etc.). This model will need to take into account dynamic of platform renewal and technology and markets emergence in the course of time. Introducing important strategies of risk management in highly uncertain situations and conditions of their implementation, we have not yet taken into account dynamics in different sectors. The further work relative to technology transfers and market emergence will need to investigate which strategy one has to adapt according to the markets and technologies dynamics. The model will integrate the emergence of new markets in the form of new functions combination and modeling of the existing and emerging technologies in the form of graph-related entities. In the proposed work, we did not take into account the competition dynamics in between different markets and other industrial actors. More investigation is necessary to understand competition dynamics influence and we will attempt to take it into account further.

Despite of the simplified model character, this work allowed determining the situations when common unknown strategy that is usually expensive, risky, and hard to manage is attractive.

Though revealing meaningful aspects of platform emergence management in highly uncertain situations, the managerial implications are still tentative due to a study limit to a single case study. Thus, the observed phenomena are proved for a success of just a semiconductor industry and more particularly one firm. Therefore, further work of model implementation to other industrial cases is highly appropriate.

## References

1. Gawer A (2009) Platforms, markets and innovation. Edward Elgar Publishing, Inc., Cheltenham
2. Baldwin YC (2008) Where do transactions come from? modularity, transactions, and the boundaries of firms. Ind Corp Change 17(1):155–195
3. Baldwin YC, Clark BK (1997) Managing in an age of modularity. Harvard Bus Rev 75(5):84–93 (Sept–Oct)
4. Gawer A, Cusumano MA (2008) How companies become platfrom leaders. MIT Sloan Manag Rev 49(2):28–35
5. Sanchez R (1996) Strategic product creation: managing new interactions of technology, markets, and organizations. Eur Manag J 14(2):121–138
6. Meyer MH, Utterback JM (1993) The product family and the dynamics of core capability. MIT Sloan Manag Rev 40(2):29–47
7. Meyer MH, Lehnerd AP (1997) The power of product platforms: building value and cost leadership. Free Press, New York
8. Krishnan V, Gupta S (2001) Appropriateness and impact of platform-based product development. Manag Sci 47(1):52–68
9. De Meyer A, Loch HC, Pitch TM (2002) Managing project uncertainty: From variation to chaos. MIT Sloan Manag Rev 43(2):60–67
10. Keenan T (2003) Identifying emerging generic technologies at the national level: the UK experience. J Forecast 22(2–3):129–160
11. Maine E, Garsney E (2006) Commercializing generic technology: the case of advanced material ventures. Res Policy 35(3):375–393
12. Utterback J (1994) Mastering the dynamics of innovation. Harvard Business School Press, Boston
13. McGrapth ME (1995) Product strategy for high-technologies companies. Irwin Professional Publishing, New York
14. Le Masson P, Hatchuel A, Weil B (2009) Platforms for the design of platforms: collaborating in the unknown. In: Gawer A (ed) Platforms, markets and innovation. Edwar Elgar Publishing, Cheltenham
15. Meyer MH, Dalal D (2002) Managing platform architectures and manufacturing processes for non assembled products. J Prod Innov Manag 19(4):277–293
16. Baldwin YC, Clark BK (2006) Modularity in the design of complex engineering systems. In: Minai A, Braha D, Yan YB (eds) complex engineering systems: science meets technology. New england complex system institute series on complexity, Springer, New York
17. Sanderson S, Uzumeri M (1995) Managing product families: the case of Sony Walkman. Res Policy 24:761–782
18. Kokshagina O, Le Masson P, Weil B, Cogez P (2012) Risk management strategies in a highly uncertain environment: understanding the role of common unknown. In: Proceedings, the 19th international product development management conference
19. Griffin RW (2004) Management. Houghton Mifflin Company, New Delhi

20. Agbadudu AB (1996) Elementary operations research, vol 1. Mudiaga Press, Benin City
21. Loch CL, De Meyer A, Pitch MT (2006) Managing the unknown: a new approach to managing high uncertainty in projects. Wiley, New York
22. Sanchez RB, Bourgault M, Pellerin R (2009) Risk management applied to projects, programs, and portfolios. Int J Managing Projects Bus 2(1):14–35
23. Chapman C (1990) A risk engineering approach to project risk management. Int J Project Manag 8:5–16
24. Petit Y, Hobbs B (2011) Project portfolio in dynamic environments: sources of uncertainty and sensing mechanisms. Project Manag J 41(4):46–58
25. Chapman C, Ward S (2003) Project risk management—process, techniques and insights, 2nd edn. Wiley, New York
26. Gawer A (2010) The organization of technological platforms. In: Philips N, Sewell G, Griffiths D (eds) Technology and organization: essays in honour of joan woodward (research in the sociology of organizations, vol 29). Emerald Group Publishing Limited, pp 287–296
27. O'Connor CG, Ravichandran T, Robenson D (2008) Risk management through learning: management practices for radical innovation success. J High Technol Manag Res 19:70–82
28. Loch CL, Solt ME, Bailey EM (2008) Diagnosing unforeseeable uncertainty in a new venture. J Prod Innov Manag 25:28–46
29. Pich TM, Loch HC, De Meyer A (2002) On uncertainty, ambiguity and complexity in project management. Manag Sci 48(8):1008–1023
30. Van de Ven HA, Polley D (1992) Learning while innovating. Organ Sci 3(1):92–116
31. Lynn SG, Akgun EA (1998) Innovation strategies under uncertainty: a contingency approach for new product development. Eng Manag J 10(3):11–17
32. Lenfle S (2011) The strategy of parallel approaches in projects with unforeseeable uncertainty: the Manhattan case in retrospect. Int J Project Manag 29:359–373
33. McGrath GR (2001) Explorative learning, innovating capacity and management oversight. Acad Manag J 44(1):118–131
34. Maidique M, Zirger B (1984) A study of success and failure in product innovation: the case of the U.S. electronics industry. IEEE Trans Eng Manag 31:192–203
35. Abernathy WJ, Clark KB (1985) Innovation: mapping the winds of creative destruction. Res Policy 14(1):3–22
36. Oriani R, Sobrero M (2001) Market valuation of firms' technological knowledge: a real options perspective. In: Extended abstract 21st annual international conference of strategic management society 2001
37. Ulrich K (1995) The role of product architecture in the manufacturing firm. Res Policy 24:419–440
38. Muffato M, Roveda M (2002) Product architecture and platforms: a conceptual framework. Int J Technol Manag 24(1):1–16
39. Joseph I (2010) Project decisions under uncertainty: applications to publicly financed project. Eur J Econ Finan Adm Sci 27:94–110
40. Lancaster Kelvin J (1966) A new approach to consumer theory. J Polit Econ 74(2):132–157
41. Benaroch M (2002) Managing information technology investment risk: a real options perspective. J Manag Inf Syst 19(2):43–84
42. Cohen MW, Levinthal DA (1989) Innovation and learning: the two faces of R&D. Econ J 99(397):569–596
43. Le Masson P, Hatchuel A, Weil B (2010) Modeling novelty-driven industrial dynamics with design functions: understanding the role of learning from the unknown. In: proceedings, the Schumpeter society conference on innovation, organization, sustainability and crises, Aalborg, 21–24 June 2010
44. Le Masson P, Cogez P, Felk Y, Weil B (2012) Revisiting absorptive capacity with a design perspective. Int J Knowl Manag Stud (Forthcoming)
45. Moore GE (1965) Cramming more components onto integrated circuits. Electron Mag 4
46. International-Roadmap-Committee (2007) ITRS—executive summary final draft. International technology roadmap for semiconductors

47. Pettigrew AM (1990) Longitudinal field research on change: theory and practice. Organ Sci 3(1):267–292
48. Chantre A, Chevalier P, Lacave T, Avenier G, Buczko M, Campidelli Y, Depoyan L, Berthier L, Gacquière C (2010) Pushing conventional SiGe HBT technology towards "Dotfive" terahertz. EuMA
49. Chevalier P (2007) High-Speed SiGe BiCMOS technologies: 120 nm status and end-of-roadmap challenges. In: Proceedings, silicon monolithic integrated circuits in RF systems, pp 18–23
50. Geynet B (2008) Développement et étude de transistors bipolaires a hétérojonctions Si/SiGe:C pour les technologies BiCMOS millimétriques. Thèse présentée a l'IEMN de Lille, France
51. Schaefer S (1999) Product design partitions with complementary components. J Econ Behav Organ 38:311–330
52. Ernst D (2005) Limits to modularity: reflections on recent development on chip design. Ind Innov 12(3):303–335
53. Ferguson CH, Morris CR (1993) Computer wars: how the west can win in a post-IBM world. Times Books, New York
54. Jacobides MG, Knudsen T, Augier M (2006) Benefiting from innovation: value creation, value appropriation and the role of industry architectures. Res Policy 35:1200–1221

# Chapter 7
# Imovator's Dilemma: How to Decide When to be Offensive and When to be Defensive?

**Türkay Dereli and Koray Altun**

**Abstract** Although being an *innovator* has some merits, being an *imitator* has its own merits as well. Staying competitive therefore requires consideration of imovation—both innovation and imitation simultaneously. An *imovator* should play role of *offensive imovator* (near to being pure innovator) when innovation is worthwhile. In case of otherwise, being a *defensive imovator* (near to being *early* imitator) can be compatible strategy. How to decide and when to be *offensive/defensive* greatly depends on *innovation potentials* of industries assessed by consideration of both their *technological* and *market conditions*. This chapter introduces a unique framework called as *quick innovation intelligence process* for assessment of *innovation potential*. It makes use of databases of patent and publication and some marketing indicators/determinants in order to take technological and market conditions into account. The framework uses *interval type-2 fuzzy sets and systems (IT2FSSs)* and a *data-fusion methodology* to combine above-mentioned databases in order to infer about innovation potential. An imovator can be aware of which strategy is well suited and can go through calculations about when innovation is worthwhile by taking likelihood of being imitated into account as well through the presented assessment process. For that reason, the process presented in this chapter can contribute to solution of the *imovator's dilemma*.

T. Dereli · K. Altun (✉)
Department of Industrial Engineering, Gaziantep University, 27310 Gaziantep, Turkey
e-mail: kaltun@gantep.edu.tr

T. Dereli
e-mail: dereli@gantep.edu.tr

D. Cetindamar et al. (eds.), *Strategic Planning Decisions in the High Tech Industry*, DOI: 10.1007/978-1-4471-4887-6_7,

## 7.1 Is Innovation Crucial to Survival?

Popularity of the term innovation has grown exponentially over the last decades. Loads of researchers and practitioners have dedicated themselves to studying management of innovation since they believe that "innovation is crucial to survival". The usage of this sentence has served as a vital base for constructing the efforts devoted to management of innovation, and therefore numerous studies related with innovation have started their discussions by using this cliché sentence. We also agree with that innovation is an important issue to remain competitive and to survive in today's business world. However, it should not be ignored that innovation creates temporary monopolies, allowing abnormal profits that would soon be competed away by rivals and imitators [1]. Besides, achieving innovation is burdened by two sources of uncertainty: First, the time span between investing is realized and its financial return is obtained, and second because it could be easily copied without incurring in the cost of research and development (R&D) [2].

Making and leading the innovation wind rather than being possessed by it actually provides benefits of monopoly power. However, investing heavily in R&D with the intention of being innovator does not warrant success and its sustainability. Although being an innovator has some merits, being an imitator has its own merits as well. There is a great attention devoted to these merits in the literature in order to find out which one is the best strategy being an innovator or an imitator (e.g., [3–7]). Sometimes being an innovator is the best strategy while introducing a product, sometimes being an imitator. Imitation is as important as innovation in reality even if innovation is associated with good perceptions. Product development can take a mixed form between two extremes on a continuum, from brand new innovation to pure imitation. According to [8], consideration of imovation–both innovation and imitation simultaneously—is required in order to combine merits of them and hence to stay competitive. Figure 7.1 shows fuzzy sets analogy for product introduction strategy. Playing offense and defense is regarded as one of the ten rules of imovation stated by [8]. An imovator should play role of *offensive imovator* (near to being pure innovator) when innovation is worthwhile. In the case of otherwise, being a *defensive imovator* (near to being *early* imitator) can be compatible strategy.

How to decide when to be *offensive/defensive* greatly depends on *innovation potentials* of industries assessed by consideration of both their *technological* and *market conditions*. High innovation potential can mean that making innovation is worthwhile, and vice versa. Evaluation of innovation potential can guide imovators to conclude when to be offensive and when to be defensive. This chapter presents a process for assessment of *innovation potential* in order to solve above-mentioned imovator's dilemma. The process presented in this chapter makes use of databases of patent and publication and some marketing indicators/determinants in order to take technological and market conditions into account. An imovator can be aware of which strategy is well suited and can go through calculations about when innovation is worthwhile by taking likelihood of being imitated into account as well through the presented assessment process.

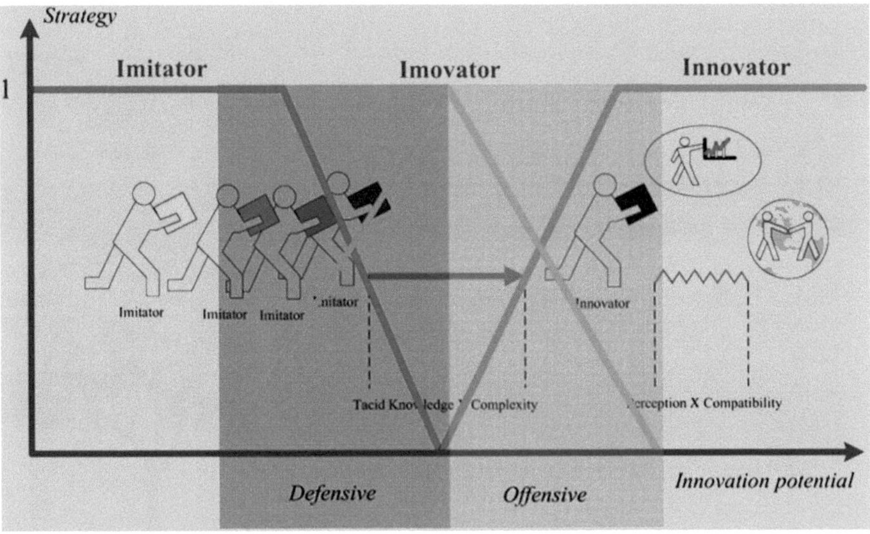


**Fig. 7.1** A suitable analogy for product introduction strategy through fuzzy sets

The reminder of this chapter is organized as follows: Section 7.2 addresses assessment of innovation potential. Section 7.3 presents a novel assessment process namely; Quick innovation intelligence process (QIIP). An application of QIIP is presented in Sect. 7.4. Finally, concluding remarks are given.

## 7.2 Assessment of Innovation Potential

### *7.2.1 Innovation Process*

Understanding essentials and life cycle of a typical innovation process is influential for developing proper frameworks that aim to improve success on product introductions. Technological *innovation process* may refer to transformation of a *technology* from an idea to a commercial product as indicated in Fig. 7.2. It has three major phases that have to be carried out sequentially as *fuzzy front end (FFE)*, *new product development (NPD)*, and *commercialization*. FFE is idea generation phase that initiates the process. The major outcomes of this phase are a well-defined concept, clear development requirements, and a business plan aligned with the corporate strategy [9]. The second phase aims to produce concrete products as outcomes. Commercialization is the last step of the innovation process. Without achieving commercialization phase, the outcome of the process may be called as *invention*. The market determines whether an invention becomes an innovation (Kusiak [10]). Having commercialized, the process continues with adoption of introduced innovation. However, from companies' perspective the process ends with (duplicative and/or creative) *imitation* by competitors.

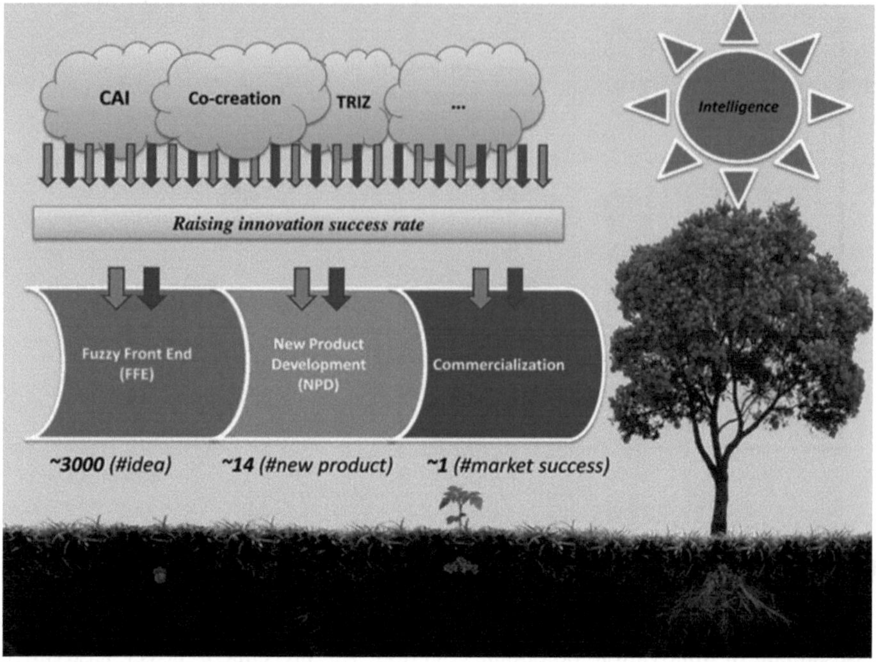


**Fig. 7.2** Innovation process

Innovation is a risky and expensive endeavor which results in low success rates (Cormican and O'Sullivan [11]). For this reason, focusing efficiency and effectiveness on innovation process is intrinsically inevitable to raise the success obtained through innovation. There have been great efforts for developing and providing appropriate infrastructures, support systems, and intelligent approaches in order to improve efficiency and effectiveness of the innovation process. In this regard, computer aided innovation (CAI) is a domain that aims to provide computerized supports to its users throughout the innovation process. Management of innovation process is getting easier and providing better solutions is being possible with computer-based applications. Especially, the theory of inventive problem solving (TRIZ) based CAI tools have been developed and help users solve the ill-structured problems. Customer co-creation and collaborative innovation in multi-site enterprises are supported by means of suitable automated negotiation mechanisms (e.g., [12, 13]).

These efforts improve the product performances especially in complex product design environments through doing things right. Nevertheless, doing the right things, i.e., effectiveness is as important as efficiency on innovation process. Effectiveness on innovation process is associated with strategic decisions in general. Assessment of *innovation potential* can contribute to make such decisions.

There have been some preliminary attempts to assess the innovation potential although the corresponding literature is scarce. Justel et al. [14] proposed a method for assessment of innovation potential of product concepts and selecting those with greater probability of success. The proposed method considers the degree of novelty of product concepts and their potential for success in the market. In another study, Jayanthi et al. [15] presented a data envelopment analysis (DEA) based application of assessment of innovation potential to a US photovoltaic industry. In the proposed approach, innovation potential is measured in terms of its relative efficiency with respect to a best practices frontier. Gupta et al. [16] proposed a method to decide innovation potential and type of innovation in a proposed design concept of a product.

Having reviewed the literature, it is observed that *commercialization* and *trendiness* factors have been taken most separately but rarely together into account. However, *imitation* has not been taken into account by the researchers for assessment of innovation potential. In the following section, a novel approach for assessment of candidate technologies with respect to their innovation potentials is presented. According to authors' best knowledge, the approach presented in this chapter is the first attempt in the literature that is dedicated to *winning innovation intelligence*—a vital perspective for solution of imovator's dilemma—and considers commercialization, trendiness and imitation factors all together.

### 7.2.2 Winning Innovation Intelligence

Innovation intelligence can be thought as a collection of intelligent approaches for assessment of technologies according to their innovation potential in order to make maximum profits. It is about the provision of relevant information on innovations and the evaluation of their impact on the corporation [17]. The goal of innovation intelligence is identification, qualification, and evaluation of technologies so as to develop a viable innovation strategy [17]. As it is discussed in the previous section, innovation creates temporary monopolies, allowing abnormal profits that would soon be competed away by rivals and imitators. Therefore, innovation potential of technologies should be evaluated with consideration of their *ability for commercialization* and their *resistance for imitation.* Such an assessment process developed for innovation potential can generate winning innovation intelligence.

## 7.3 Quick Innovation Intelligence Process

Quick innovation intelligence process (QIIP) is a process developed for the assessment of innovation potential of candidate technologies. It is based on innovation intelligence perspective and dedicated to achieve winning innovation intelligence. QIIP is first introduced by Dereli and Altun [18]. Achieving winning innovation intelligence requires commercial potential evaluation and imitation potential evaluation in order to determine ability for commercialization and resistance for imitation. However, QIIP takes trendiness of technologies into account as well to be quick on the evaluation process and to eliminate dated technologies before assessment of their innovation potentials. This helps to make quick response to owners of the candidate technologies waiting investment/ incentive decision, although they have not a worthwhile technology in their hands. Thus, QIIP is designed as it has four main steps for assessment of technologies in accordance with their innovation potential as follows:

Step 1. Trendiness evaluation
Step 2. Evaluation of commercial potential
Step 3. Evaluation of imitation potential
Step 4. Grouping and prioritizing of candidate technologies.

### *7.3.1 Trendiness Evaluation*

Evaluation and consideration of trendiness of candidate technologies is one of the most important prerequisites in order to make rational investment decisions, to draw strategic roadmaps and to direct investments and incentives to the most rewarding technologies. Scientific and technological findings are generally transformed into publications, patents, and/or industrial applications in due course. This transformation process is quite essential in order to introduce, protect, and commercialize those findings. Therefore, patents and publications have been used as a vital source to evaluate the status of science and technology. In this regard, Scientometrics (Bibliometrics) is a research method focused frequency-based quantitative exploration of publications. Technometrics takes place instead of Scientometrics when patent data are explored.

Once a specific technological thrust has been identified, Scientometrics and Technometrics can be used to determine its position in its life cycle [19]. For this reason, many researchers have used data derived from publications and patents in order to evaluate science and technology [20]. For instance, Lee et al. [21] used data derived from patents for modeling trends and patterns of innovation in energy sector. Daim et al. [22] forecasted some emerging technology areas through integrating the use of bibliometrics and patent analysis into scenario planning, growth curves, and analogies. In another study, Yu and Lo [23] developed a fuzzy inference system (FIS) for technological strategy planning by the help of using only patent data as input of the system. Huang and Li [24] proposed a framework based on time series analysis, patent analysis, and patent international-patent-classification (IPC) analysis in order to evaluate technology trend.

Scientometrics and Technometrics are well-established methods in the evaluation of science and technology. Any attempt to match the existing metrics to the evaluation scheme would almost encounter gaps, challenges, and unanswered questions [25]. Therefore, while performing strategic technology evaluation, data derived from patents and publications should be used together as matching each other in order to direct the considerations to use-inspired basic researches, and hence achieve science-based technological improvements. There is nevertheless a need to consider the linkages between the conceptual background of scientific generation and progress—and the measurement of its process and outcomes [26].

However, limited study addresses matching data of patents and publications in the literature. Bengisu and Nekhili [27] used the data derived from patents and publications to quantify and test expert views on selected technologies comparing the number of patents and publications related to the same technologies for a given year. In another study, Arman et al. [20] developed a methodology that tries to combine three different rankings obtained from publications, patents, and experts opinions in order to obtain a unique ranking. In a recent study, Quintella et al. [28] presented a contextualized overview of $CO_2$ capture technology, with critical evaluation of state-of-the-art and technological development through patents applications and scientific publications. In another recent study, Zhang et al. [29] proposed a process integrating expert knowledge and bibliometric methods include terms frequency analysis and association analysis in order to engage the challenge of technology roadmapping. Their terms frequency analysis uses technology core terms and IPCs retrieved from publication and patent documents.

Even though fuzzy logic is an effective tool for matching different kinds of data sources, it has not found any application in the literature for matching publication and patent data within a technology evaluation framework. This section presents a technology evaluation framework that is based on interval type-2 fuzzy sets and systems (IT2FSSs) in order to match patent and publication data. This framework is extensively discussed and compared with their type-1 counterparts in Dereli and Altun [30]. Figure 7.3 shows general structure of the technology evaluation framework. Following subsections of this section present in details about which data sources are used, how they are derived and processed, and how this structure works, point by point.

#### 7.3.1.1 Patents and Publications are Utilized as Input

Patent counts and publication counts are the uncontaminated input sources of this technology evaluation framework. The patent count data are retrieved from the online database of European patent office (EPO). For retrieving publication count

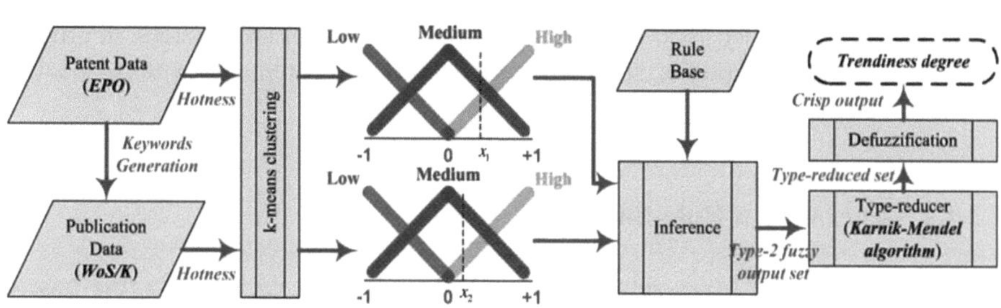


**Fig. 7.3** Technology evaluation framework that is based on type-2 FIS

data, the framework uses online database of Web of science/knowledge (WoS/K). For each candidate technology, an European classification system (ECLA) class is specified. Afterwards, some keywords are needed to be generated to make a connection between patents and their related publications. Selection of the keywords can be accomplished with the help of experts of relevant technology class. For another way, given definitions for the corresponding classes by ECLA can also be very helpful to generate keywords.

#### 7.3.1.2 "Hotness" Values are Considered to Generate Fuzzy Input Sets

Arman et al. [20] determine *hotness* values by calculating the number of patents appearing in the last 3 years as a percentage of those that appeared in the last 10 years. Use of *"hotness"* values instead of amounts is more appropriate to evaluate trendiness of technologies, since it is related with growth rate of technologies that are more suitable for trendiness evaluation. In this framework, we measure the *hotness* values in order to evaluate candidate technologies with respect to their trendiness. However, perception of the term, *hotness,* can vary from person to person. Therefore, the term *hotness* is also fuzzy. In order to handle this uncertainty, type-2 fuzzy sets are used in this framework.

#### 7.3.1.3 Membership Functions are Designed Through "k-means Clustering"

Both patent data and publication data are classified into three clusters as *low*, *medium*, and *high.* In this step, in order to find centroids of the clusters, we use a well-known technique; namely, *k-means clustering* developed by MacQueen [31]. In the first step of this technique, initial guesses are made for the means of *low*, *medium*, and *high.* These points represent initial group centroids. In the next step, every datum of patents and publications is assigned to the cluster that has the closest centroids. These two steps alternate until there are no changes in any mean. The values limiting these clusters are used to define fuzzy membership functions for a certain patent and publication data. In the design of input fuzzy sets, we use triangular membership functions, these are have been frequently used because of their striking simplicity [32]. Having found the centroids of the clusters, in order to find left and right-end points, standard deviations of the patent data and publication data can be used.

#### 7.3.1.4 Generating a Rule-Base

After developing the fuzzy sets for each input and output variables, a rule-base is needed to be generated. We use type-2 FIS for matching patents and publications. Therefore, we do not need past experience to generate a rule-base. However, we should know which data source is more reliable. The rule-base should be generated in accordance with the importance weight of patent and publication data.

#### 7.3.1.5 Inference Process

Having performed input processing step, inference process is performed as follows;

Consider a rule-base that includes $N$ rules as;

Rule (n): If $x_1$ is $\tilde{X}_1^n$ and $x_2$ is $\tilde{X}_2^n$ then $y$ is $Y^n$ $n = 1, 2, \ldots, N$, where $\tilde{X}_1^n$ are interval type-2 fuzzy sets which are generated from patent data and $\tilde{X}_2^n$ are interval type-2 fuzzy sets which are generated from publication data. $x_1$ and $x_2$ are the calculated patent and publication hotness values of candidate technologies, respectively. $Y^n$ values are intervals ($= \left[\underline{y}^n, \bar{y}^n\right]$) which represents the trendiness in a gradual manner.

Compute the membership of $x_1$ on each

$$\tilde{X}_1^n, \left[\mu_{\underline{X}_1^n}(x_1), \ \mu_{\bar{X}_1^n}(x_1)\right], \quad n = 1, 2, \ldots, N \tag{7.1}$$

Compute the membership of $x_2$ on each

$$\tilde{X}_2^n, \left[\mu_{\underline{X}_2^n}(x_2), \ \mu_{\bar{X}_2^n}(x_2)\right], \quad n = 1, 2, \ldots, N \tag{7.2}$$

Compute the firing interval of the $n$th rule, $F^n(x_1, x_2)$, through following equation.

$$F^n(x_1, x_2) = \left[\mu_{\underline{X}_1^n}(x_1) \times \mu_{\underline{X}_2^n}(x_2), \mu_{\overline{X}_1^n}(x_1) \times \mu_{\overline{X}_2^n}(x_2)\right] \equiv \left[\underline{f}^n, \overline{f}^n\right], \quad n = 1, 2, \ldots, N \tag{7.3}$$

#### 7.3.1.6 Type-Reduction Process

Type-reduction process aims to convert type-2 fuzzy sets into type-1 fuzzy sets for preparation to defuzzification process. Center of sets ($Y_{\cos}$) type reducer, that is one of the most commonly preferred type reducers, is used to perform the type-reduction process. $Y_{\cos}$ is expressed as following equation.

$$Y_{\cos}(x) = \bigcup_{\substack{f^n \in F^n(x) \\ y^n n}} \frac{\sum_{n=1}^{N} f^n y^n}{\sum_{n=1}^{N} f^n} = [y_l, y_r] \tag{7.4}$$

$y_l$ and $y_r$ are the end points of the interval set. They are expressed as following equations, respectively.

$$y_l = \frac{\sum_{n=1}^{L} \bar{f}^n \underline{y}^n + \sum_{n=L+1}^{N} \underline{f}^n \underline{y}^n}{\sum_{n=1}^{L} \bar{f}^n + \sum_{n=L+1}^{N} \underline{f}^n} \tag{7.5}$$

$$y_r = \frac{\sum_{n=1}^{R} \underline{f}^n \bar{y}^n + \sum_{n=R+1}^{N} \bar{f}^n \bar{y}^n}{\sum_{n=1}^{R} \underline{f}^n + \sum_{n=R+1}^{N} \bar{f}^n}$$

where switch points $L$ and $R$ are specified by

$$\underline{y}^L \leq y_l \leq \underline{y}^{L+1} \text{ and } \bar{y}^R \leq y_r \leq \bar{y}^{R+1} \tag{7.6}$$

Karnik–Mendel (KM) algorithm, which is one of the most common approaches in the literature, is employed in order to find switch points for each end points of the interval set. The steps of the algorithm are as follows [33]; KM algorithm for computing $y_l$:

Step 1: $\underline{y}_n$ values are sorted in increasing order.
Step 2: The weights $F^n(x)$ are matched with their respective $\underline{y}_n$ values.
Step 3: $f^n$ are initialized through $f^n = \frac{\underline{f}^n + \bar{f}^n}{2}$ and then $y$ is computed as;

$$y = \frac{\sum_{n=1}^{N} \underline{y}^n f^n}{\sum_{n=1}^{N} f^n} \tag{7.7}$$

Step 4: Switch point $k (1 \leq k \leq N-1)$ is found as $\underline{y}^k \leq y \leq \underline{y}^{k+1}$.
Step 5: $f^n = \begin{cases} \bar{f}^n, n \leq k \\ \underline{f}^n, n > k \end{cases}$ are set and then $y'$ is computed as;

$$y' = \frac{\sum_{n=1}^{N} \bar{y}^n f^n}{\sum_{n=1}^{N} f^n} \tag{7.8}$$

Step 6: Check if $y' = y$. If yes, stop and set $y_l = y$ and $L = k$.

If no, go to Step 7.

Step 7: Set $y = y'$ and go to Step 4.

#### 7.3.1.7 KM Algorithm for Computing $\mathbf{y_r}$

Step 1: $\bar{y}_n$ values are sorted in increasing order.
Step 2: The weights $F^n(x)$ are matched with their respective $\bar{y}_n$ values.
Step 3: $f^n$ are initialized through $f^n = \frac{\underline{f}^n + \bar{f}^n}{2}$ and then y is computed as;

$$y = \frac{\sum_{n=1}^{N} \bar{y}^n f^n}{\sum_{n=1}^{N} f^n} \tag{7.9}$$

Step 4: Switch point $k(1 \leq k \leq N-1)$ is found as $\bar{y}^k \leq y \leq \bar{y}^{k+1}$.

Step 5: $f^n = \begin{cases} f^n, n \leq k \\ \bar{\bar{f}}^n, n > k \end{cases}$ are set and then $y'$ is computed as;

$$y' = \frac{\sum_{n=1}^{N} \bar{y}^n f^n}{\sum_{n=1}^{N} f^n} \tag{7.10}$$

Step 6: Check if $y' = y$. If yes, stop and set $y_r = y$ and $R = k$.

If no, go to Step 7.

Step 7: Set $y = y'$ and go to Step 4.

#### 7.3.1.8 Defuzzification Process

Having performed KM algorithm, we reach to value of switch points of the interval set. Afterwards, defuzzified output is computed using following equation.

$$y = \frac{y_l + y_r}{2} \tag{7.11}$$

After defuzzification process, we obtain a crisp value for each candidate technology which shows the *trendiness* of corresponding technology through fusing patent data and publication data. These values are then used to rank the candidate technologies with respect to their *trendiness*.

### 7.3.2 Evaluation of Commercial Potential

Çetindamar et al. [34] highlight that many companies are good at producing inventions but not all their inventions are put into use. Commercial potential of candidate technologies needs to be evaluated in order to improve this success rate. Various evaluation methods have been developed in the literature. See [35, 36] to find reviews on corresponding evaluation methods.

Strategic technology evaluation program (STEP) is the most common tool for evaluation of new technologies at the early stages of technology development [37]. STEP method is based on opinion on experts. Requirement for an expert prevents computerizing the evaluation process. However, the QIIP is designed for being infrastructure of a CAI tool. The proposed process therefore should use determinants/indicators of commercial potential rather than opinion of experts. With this purpose in mind, it is preferred to use the results of market researches as commercial potential determinants/indicators. Market research monitor (www.researchmonitor.euromonitor.com) is a brand new online collection of huge amounts of market reports. Designed specifically for libraries, users gain a complete view of the consumer marketplace in 80 countries. It enables its users to access professional and high quality market data, to identify leading subcategories in each market, to know 5-year forecasts to assess how each category will develop. QIIP evaluates commercial potentials by ranking corresponding markets in accordance with their percentage of expected average growth (PEAG).

### 7.3.3 Evaluation of Imitation Potential

Imitation sabotages the innovation process. The innovator firms lose their monopoly power and market shares of them get smaller. For that reason, imitation plays and important role for firms. It has to be considered while doing strategic plans. In the literature, although protection of innovation has been intensively studied (see [38–40]), evaluation of imitation potential of candidate technologies in the early stage has hardly ever studied.

It is asserted that tacit knowledge and complexity are especially important to prevent imitating the innovation [41]. Tacit knowledge is noncodified, disembodied know-how that is acquired via the informal take-up of learned behavior and procedures [42]. Imitation is blocked through tacit knowledge, because it cannot be directly or easily transmitted by codifying and learning. Therefore, it can help innovator firms to resist imitation by competitors.

How can we realize in the early stage whether a technology includes tacit knowledge or not? Competitive landscape of the candidate technologies can help us to infer about imitation potential. Market Research Monitor can be helpful for identifying the imitation potential as well. The information of top players and their brands, and assessment company shares in the corresponding market can give an idea about the degree of tacitness. QIIP uses ranking list obtained by consideration of the average market shares of top players of the corresponding candidate technologies.

### 7.3.4 Grouping and Prioritizing of Candidate Technologies

Having performed the evaluation steps, each evaluation gives a ranking for candidate technologies. However, a unique ranking list is needed at the end of the process. A data fusion operation is therefore required to combine these three different ranking lists. The problem faced in this stage is that there is no explicit correlation among results of the three evaluations. For this reason, data fusion to make a unique ranking list of candidate technologies may be questionable. In their study, Arman et al. [20] used a simple prioritizing method as indicated in Fig. 7.4 for the same situation which there is no explicit correlation between data sources. With the method, rather than ranking the technologies precisely, a grouping is achieved in order to emphasize the importance of candidate technologies. QIIP uses the same method in order to make grouping and prioritizing of candidate technologies according to their innovation potential.

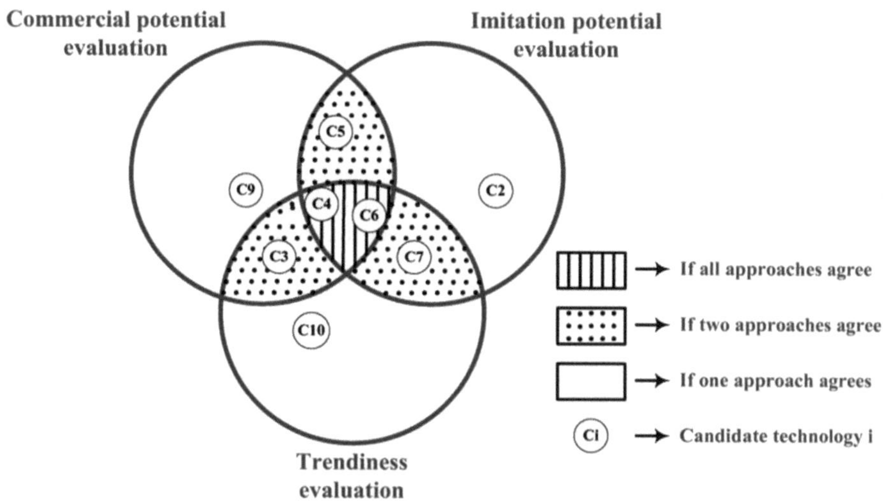


| | |
|---|---|
| A | C4, C6 |
| B | C3, C5, C7 |
| C | C2, C9, C10 |
| D | C1, C8 |

**Fig. 7.4** Grouping and prioritizing candidate technologies

The grouping method works as follows: assume that we have 10 candidate technologies and evaluated these technologies according to their commercial potential, imitation potential, and trendiness. Three different ranking lists we will have after the evaluations. At this stage, we will specify a limit according to budget which will be used for investment incentives. Assume that top five technologies for each list are determined. If a candidate technology is in the top five of the every list, this technology will be member of the class A. If the technology is in the top five of any two of lists, this technology will be member of the class B. And the rest of them will be the member of class C. Beside the candidate technologies which they are not in top five in any list will be the member of class D. In the ending, QIIP enables us to group and prioritize candidate technologies according to their innovation potentials.

## 7.4 An Application

An application of QIIP is presented in this section. Following candidate technologies are evaluated in accordance with their innovation potentials (see Table 7.1).

**Table 7.1** The candidate technologies

| Candidates | Market | ECLA |
|---|---|---|
| C1 | Camcorders | H01L |
| C2 | Cameras | H01P |
| C3 | Computers and peripherals | H01S |
| C4 | Home audio and cinema | H01G |
| C5 | In-car-entertainment | H01B |
| C6 | Mobile phones | H01T |
| C7 | Portable media players | H01Q |
| C8 | Televisions and projectors | H01R |
| C9 | Video players | H01C |

**Table 7.2** Center values of the MFs

| MFs | | Patents | Publications |
|---|---|---|---|
| Hotness1 | High | 0,2423 | 0,3827 |
| | Medium | 0,1921 | 0,3040 |
| | Low | 0,1514 | 0,2391 |
| Hotness2 | High | 0,3474 | 0,4538 |
| | Medium | 0,2867 | 0,4000 |
| | Low | 0,2360 | 0,3448 |
| Hotness3 | High | 0,4547 | 0,5497 |
| | Medium | 0,3833 | 0,4837 |
| | Low | 0,3302 | 0,4349 |

For pre-elimination and trendiness evaluation of candidate technologies, initially, patents count data are retrieved from the database of EPO. Three types of hotness values are used to generate type-2 fuzzy sets. For *hotness1*, last 2 years divided by last 10 years. For *hotness2*, last 3 years divided by last 10 years. For *hotness3* values, last 5 years divided by last 10 years.

The next step is to generate some keywords to make a link between patents and publications. The keywords are obtained from our earlier study [43]. They are determined by using the definitions of the each subclass. Afterward, the numbers of publications are obtained by using the database of WoS/K. Three types of hotness values are calculated by using the data for publications as well. Calculated hotness values are classified into three groups as low, medium, and high for both patents and publications through the use of k-means clustering technique. Table 7.2 shows the center values of the MFs. Three input type-1 fuzzy sets are obtained for each hotness values. Through observing LMFs and UMFs for each linguistic variable, type-2 fuzzy input sets are obtained as shown in Fig. 7.5.

Verbeek et al. [44] paid attention to that some technology fields are highly science oriented while others are not. There are not equal likelihoods for different patent classes to link to science that is published in peer reviewed journals [45]. Therefore, science orientation of corresponding patent classes also need to be taken into account while generating a rule-base since reliability of publications and patents can change with respect to different patent classes. See Boyack and

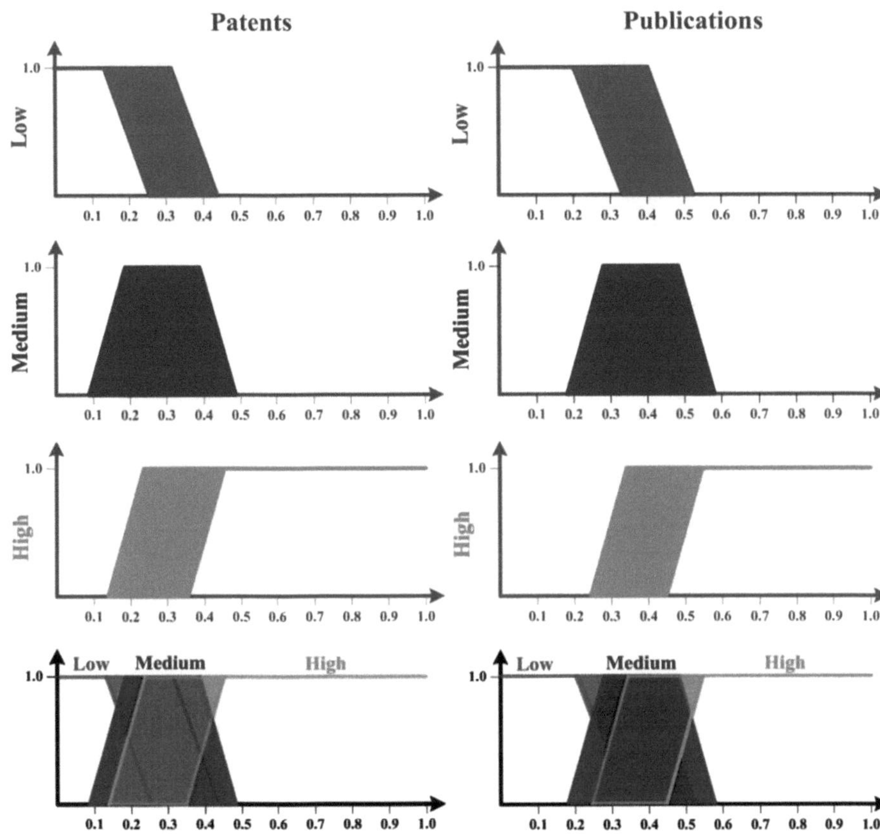


**Fig. 7.5** Type-2 fuzzy sets for patents and publications

Klavans [45] in order to review a map of IPC patent subclasses that include the distribution of patent classes with high science orientation and low science orientation. H01-Basic Electric Elements class is one of the highest science-oriented patent classes. Therefore, by taking into account this property provided by H01-Basic Electric Elements class, we generate a rule-base that has homogenous consequence distribution as shown in Table 7.3.

Average hotness values of each technology class are the inputs of the inference system. The firing intervals of the nine rules are calculated for each candidate technology. In the type-reduction process, KM algorithm is executed to find the switch points for each end points of the interval set of each candidate technology. After finding the end points, defuzzification process is executed. For each candidate technology, type-2 FIS provides a crisp output which represents the trendiness degree that is obtained by matching the data of corresponding patents and publications. Table 7.4 shows the results of trendiness evaluation of candidate technologies.

**Table 7.3** Rule-base

| | |
|---|---|
| R1 | If (Avrg *Hotness* of Patent is *Low*) and (Avrg *Hotness* of Publication is *Low*) then (Trendiness of Technology is [0.1,0.4]) |
| R2 | If (Avrg *Hotness* of Patent is *Low*) and (Avrg *Hotness* of Publication is *Medium*) then (Trendiness of Technology is [0.1,0.4]) |
| R3 | If (Avrg *Hotness* of Patent is *Low*) and (Avrg *Hotness* of Publication is *High*) then (Trendiness of Technology is [0.4,0.6]) |
| R4 | If (Avrg *Hotness* of Patent is *Medium*) and (Avrg *Hotness* of Publication is *Low*) then (Trendiness of Technology is [0.1,0.4]) |
| R5 | If (Avrg *Hotness* of Patent is *Medium*) and (Avrg *Hotness* of Publication is *Medium*) then (Trendiness of Technology is [0.4,0.6]) |
| R6 | If (Avrg *Hotness* of Patent is *Medium*) and (Avrg *Hotness* of Publication is *High*) then (Trendiness of Technology is [0.6,0.9]) |
| R7 | If (Avrg *Hotness* of Patent is *High*) and (Avrg *Hotness* of Publication is *Low*) then (Trendiness of Technology is [0.4,0.6]) |
| R8 | If (Avrg *Hotness* of Patent is *High*) and Avrg *Hotness* of Publication is *Medium*) then (Trendiness of Technology is [0.6,0.9]) |
| R9 | If (Avrg *Hotness* of Patent is *High*) and (Avrg *Hotness* of Publication is *High*) then (Trendiness of Technology is [0.6,0.9]) |

**Table 7.4** Ranking list obtained from trendiness evaluation

| Technology | C1 | C2 | C3 | C4 | C5 | C6 | C7 | C8 | C9 |
|---|---|---|---|---|---|---|---|---|---|
| Trendiness | 0,606 | 0,574 | 0,570 | 0,578 | 0,623 | 0,606 | 0,614 | 0,616 | 0,616 |
| Rank | 5 | 8 | 9 | 7 | 1 | 6 | 4 | 3 | 2 |

**Table 7.5** Ranking list obtained from evaluation of commercial potential

| Technology | C1 | C2 | C3 | C4 | C5 | C6 | C7 | C8 | C9 |
|---|---|---|---|---|---|---|---|---|---|
| PEAG | 37,7 | 39 | 16,6 | 23,8 | 17,5 | 8,6 | 10 | 27 | 1 |
| Rank | 2 | 1 | 6 | 4 | 5 | 8 | 7 | 3 | 9 |

In order to evaluate the commercial potential of candidate technologies, the database of Market Research Monitor is used. It enables us to know 5-year forecasts about corresponding markets which candidate technologies will be a participant. Commercial potentials are ranked in accordance with the percentage of expected average growth (PEAG) of each market in Turkey. Table 7.5 shows the ranking list obtained from commercial potential.

While evaluating imitation potential of technology classes, we do not have information about the complexity of candidate technologies in our example. In that case, competitive landscape of the candidate technologies can help us to infer about imitation potential. Information of top players and assessment of company shares in the corresponding market can provide an idea about imitation potential. Therefore, in this stage we have used the competitive landscape data of Turkey by using database of Market Research Monitor. A ranking list is obtained in

**Table 7.6** Ranking list obtained from evaluation of imitation potential

| Technology | C1 | C2 | C3 | C4 | C5 | C6 | C7 | C8 | C9 |
|---|---|---|---|---|---|---|---|---|---|
| Rank | 4 | 6 | 3 | 5 | 7 | 1 | 9 | 2 | 8 |

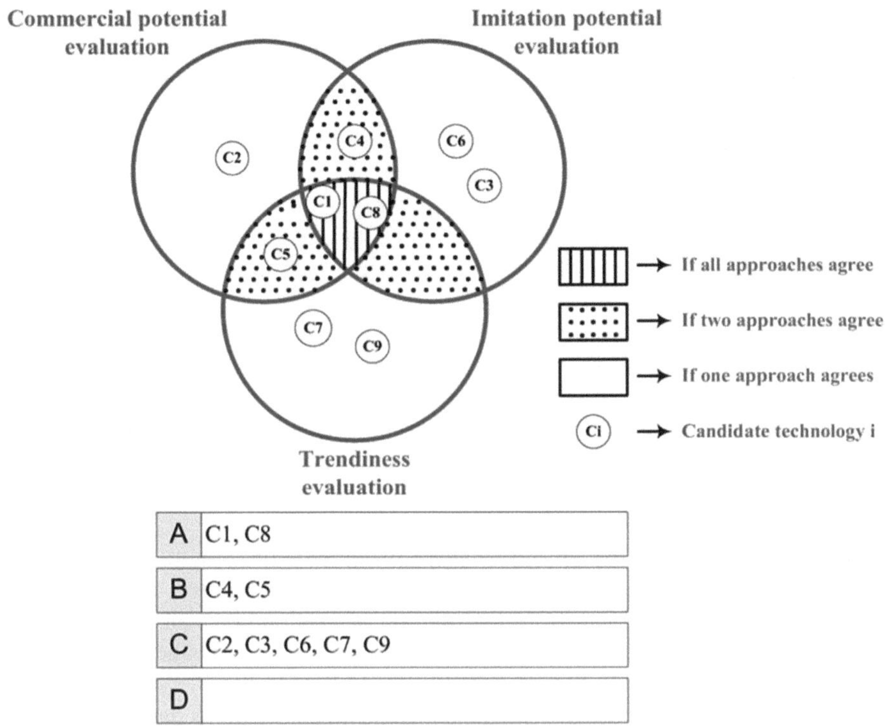


**Fig. 7.6** Grouping candidate technologies according to their innovation potentials

accordance with the average market shares of top players of the corresponding technologies. The ranking list is given in Table 7.6.

After evaluating the technologies in accordance with the three different aspects, the last step of the quick innovation intelligence process, grouping, is performed to combine the ranking lists. Figure 7.6 shows groups obtained by combining the ranking lists. Grouping shows that "C1" and "C8" are the technologies which are agreed by all approaches. "C4" and "C5" are agreed by two approaches. The rest of them are agreed by only one approach. As a result of the proposed process, the candidate technologies are classified according to their innovation potential. The output of the process can be helpful making more reliable decisions to the decision makers.

## 7.5 Concluding Remarks

Sometimes innovation is the best strategy while at other times imitation takes place instead of innovation. Fusing merits of innovation and imitation what can be termed as imovation can generate the most benefits that can be obtained in today's markets. While pursuing a strategy of imovation, a strategic decision on being offensive or defensive is needed to be made. This chapter helps imovators to make this strategic decision by presenting a novel process to assessment of innovation potential, namely: quick innovation intelligence process (QIIP). The process enables us to evaluate candidate technology areas according to their innovation potentials relatively. The output of this process is grouped and prioritized technology areas. If a candidate technology area drops into class "A" and "B" it can utilize as a sign to pursue a strategy of *higher offensive* and *offensive* imovation, respectively. Class "C" and "D" can also utilize as a sign to pursue a strategy of *defensive* and *higher defensive* imovation, respectively.

Several major advantages of QIIP can be given as follows. In the literature, although commercialization and trendiness factors have been taken mostly separately but rarely together into account, imitation has not been taken into account by the researchers for technology assessment in the early stage. QIIP considers these three factors all together. Therefore, the outcomes of the assessment results can be more beneficial for sustainable success of firms and national economies when the process is executed.

QIIP uses determinants and indicators of innovation potential rather than opinions of experts. The framework reduces the overall duration of evaluations. Moreover, being independent from the opinion of experts gives an opportunity to develop computer-aided tools. Therefore, QIIP can also be considered within the scope of CAI methods.

In the trendiness evaluation step, patents data and publications data are matched through an interval type-2 fuzzy inference system. Although fuzzy logic is an effective tool for fusing different kinds of data sources, it has not found any application in the literature for fusing publication and patent data within a technology evaluation framework. The work presented in this study is also the first attempt in the literature that proposes a trendiness evaluation that is based on fuzzy logic.

Making comparative analysis of the existing industrial regions through the proposed assessment process and generating strategic roadmaps should also be addressed in future studies.

## 7.6 Contribution and Suggestion to Managers

It has been entirely discussed in the literature that innovation has some merits but imitation has its own merits as well. Rather than dedicating R&D efforts to solely innovation or imitation, fusing together of innovation and imitation (named as imovation) is decisive to make synergy on R&D activities by benefiting from their merits simultaneously. While pursuing a strategy of imovation, a decision on being offensive or defensive is intrinsically required. This study is the first attempt in the literature to solve this dilemma. In order to solve this dilemma, this chapter presents a process for assessment of technology areas in accordance with their innovation potentials. The higher innovation potential means that making innovation is worthwhile, and vice versa. Therefore, assessment of innovation potential can be used as the signs of being an offensive or defensive imovator.

This chapter presents an assessment process, namely: quick innovation intelligence process. QIIP makes use of interval type-2 fuzzy sets and systems and a data-fusion methodology to combine patent data, publication data, and market reports to infer about innovation potentials. According to our best knowledge, this is the first attempt in the literature that is a use of type-2 fuzzy sets and systems to make strategic decisions in the high-tech industries and within a technology evaluation framework.

## References


1. Pol E, Carroll P (2006) An introduction to economics with emphasis on innovation. Thomson, MelbourneEscribano A, Giarratana MS (2011) EU patent system: to be or not to be? Working paper 11–01, Economic Series
3. Lieberman MB, Montgomery DB (1988) First-mover advantages. Strateg Manag J 9(1–2):41–58
4. Bowman D, Gatignon J (1996) Order of entry as a moderator of the effect of the marketing mix on market share. Marketing Science 15(3):222–242
5. Robinson WT, Min S (2002) Is the first to market the first to fail? Empirical evidence for industrial goods businesses. J Mark Res 34(1):120–128
6. Zhou KZ (2006) Innovation, imitation, and new product performance: the case of China. Ind Mark Manage 35:394–402
7. Chang MH, Harrington JE (2007) Innovators, imitators, and the evolving architecture of problem-solving networks. Organ Sci 18(4):648–666
8. Shenkar O (2010) Copycats: how smart companies use imitation to gain a strategic edge. Harvard Business Press, Boston
9. Kim J, Wilemon D (2002) Focusing the fuzzy front end in new product development. R&D Management 32(4):269–279
10. Kusiak A (2009) Innovation: a data-driven approach. Int J Prod Econ 12:440–448
11. Cormican K, O'Sullivan D (2004) Auditing best practice for effective product innovation management. Technovation 24:819–829
12. Dereli T, Altun K (2012) Modified Even-Swaps: a novel, clear, rational and easy-to-use mechanism for multi-issue negotiation. Comput Ind Eng 63(4):1013–1029
13. Altun K, Dereli T, Baykasoglu A (2012) Development of a framework for customer co-creation in NPD through multi-issue negotiation with issue trade-offs. Expert Syst Appl 40(3):873–880
14. Justel D, Vidal R, Arriaga E, Franco V, Val-Jauregi E (2007) Evaluation method for selecting innovative product concepts with greater potential marketing success. In: Proceedings of 16th international conference on engineering design, Paris, France
15. Jayanthi S, Witt EC, Singh V (2009) Evaluation of potential of innovations: a DEA-based application to US photovoltaic industry. IEEE Trans Eng Manage 56(3):478–493

16. Gupta PC, Garg S, Maheshwari S (2011) Evaluation and selection methodology for an innovative product design concepts. Int J Eng Sci Technol 3(4):3553–3561
17. Golovatchev J, Budde O (2010) Technology and innovation radar—Effective instruments for the development of a sustainable innovation strategy. In: Proceedings of IEEE international conference on management of innovation and technology (ICMIT), pp 760–764
18. Dereli T, Altun K (2012) A novel approach for assessment of candidate technologies with respect to their innovation potentials: quick innovation intelligence process. Expert Syst Appl 40(3):881–891
19. Martino JP (2003) A review of selected recent advances in technological forecasting. Technol Forecast Soc Chang 70:719–733
20. Arman H, Hodgson A, Gindy N (2009) Technologies watch exercise: foresight approach enhanced with scientific publications and patents analysis. Int J Technol Intell Plann 5:305–332
21. Lee S, Lee H, Yoon B (2012) Modeling and analyzing technology innovation in the energy sector: patent based HMM approach. Comput Ind Eng. doi:10.1016/j.cie.2011.12.002
22. Daim TU, Rueada G, Martin H, Gerdsri P (2006) Forecasting emerging technologies: use of bibliometrics and patent analysis. Technol Forecast Soc Chang 73:981–1012
23. Yu WD, Lo SS (2009) Patent analysis-based fuzzy inference system for technological strategy planning. Autom Constr 18(6):770–776
24. Huang L, Li Y (2010) Evaluation frame of technological developing trend based on patent information. In: Proceedings of 12th international conference on computer modelling and simulation, pp 299–304
25. Geisler E (2002) The metrics of technology evaluation: where we stand and where we should go from here. Int J Technol Manage 24(4):341–374
26. Geisler E (2005) The measurement of scientific activity: research directions in linking philosophy of science and metrics of science and technology outputs. Scientometrics 62(2):269–284
27. Bengisu M, Nekhili R (2006) Forecasting emerging technologies with the aid of science and technology databases. Technol Forecast Soc Chang 73:835–844
28. Quintella CM, Hatimondi SA, Musse APS, Miyazaki SF, Cerqueira GS, Moreira AA (2011) CO2 capture technologies: an overview with technology assessment based on patents and articles. Energy Procedia 4:2050–2057
29. Zhang Y, Zhu D, Wang X, Guo Y, Chen H, Liu C, Xu J, Heng X, Lu X (2011) Composing technology roadmapping according to bibliometrics: hybrid model and empirical study. In: Proceedings of CSIT, international conference on computer communication and management, vol 5, pp 405–410
30. Dereli T, Altun K (2013) Technology evaluation through the use of interval type-2 fuzzy sets and systems. Comput Ind Eng (in revision)
31. MacQueen J (1967) Some methods for classification and analysis of multivariate observations. In: Proceedings of 5th Berkeley symposium on mathematical statistics and probability, vol 1. University of California Press, Berkeley, pp 281–297
32. Pedrycz W (1994) Why triangular membership functions? Fuzzy Sets Syst 64:21–30
33. Mendel JM, Wu D (2010) Perceptual computing: aiding people in making subjective judgments. IEEE Press Series on Computational Intelligence, Hoboken
34. Çetindamar D, Phaal R, Probert D (2010) Technology management activities and tools. Palgrave/Macmillan, Basingstoke
35. Bandarian R (2007) Evaluation of commercial potential of a new technology at the early stage of development with fuzzy logic. J Technol Manage Innov 2:73–85
36. Sohn SY, Moon TH, Kim S (2005) Improved technology scoring model for credit guarantee fund. Expert Syst Appl 28:327–331
37. Jain RK, Martyniuk AO, Harris MM, Niemann RE, Woldmann K (2003) Evaluating the commercial potential of emerging technologies. Int J Technol Transfer Commercialisation 2:32–50

38. Lemer J (2009) The empirical impact of intellectual property rights on innovation: puzzles and clues. Am Econ Rev Papers Proc 99:343–348
39. Park W (2008) Intellectual property rights and international innovation. In: Maskus K (ed) Intellectual property, Netherlands: growth and trade. Elsevier, Netherlands
40. Qian Y (2007) Do national patent laws stimulate domestic innovation in a global patenting environment? A cross-country analysis of pharmaceutical patent protection, 1978–2002. Rev Econ Stat 89:436–453
41. Robert MG (2009) Contemporary strategy analysis. Wiley, West Sussex. ISBN 978-0-470-74710-0
42. Howels J (1996) Tacit knowledge, innovation and technology transfer. Technol Anal Strateg Manage 8:91–106
43. Dereli T, Durmusoglu A, Altun K, Bozyer Z (2010) Predicting the future of technology by using data fusion methodologies: techpredict v1.0. In: Proceedings of IMS 2010—the 7th international symposium on intelligent manufacturing systems, pp 215–223
44. Verbeek A, Debackere K, Luwel M, Andries P, Zimmermann E, Deleus F (2002) Linking science to technology: using bibliographic references in patents to build linkage schemes. Scientometrics 54(3):399–420
45. Boyack KW, Klavans R (2008) Measuring science-technology interaction using rare inventor-author names. J Informetrics 2:173–182

# Chapter 8
# Technology Acquisition by Collaboration: A Conceptual Framework

**Victor G. Ortiz-Gallardo, David Probert and Robert Phaal**

**Abstract** Collaboration has become a recurrent practice in business to acquire technology; however, it can be a risky option when the factors that may affect the outcomes are unknown. A large number of papers have explored the impact of different factors on the performance of technology acquisitions and collaborations, but no framework has been proposed that integrates all the factors that influence the acquisition of technology by means of collaboration. This chapter aims to fill this gap in the existing literature. Through an analysis of the academic literature, a conceptual framework is presented that provides an overview of the key aspects that have influence on the performance of technology acquisitions by collaboration. The evidence suggests that five types of factors are relevant: contextual issues, technology uncertainty, strategic alignment, structural match and coordination capability.

## 8.1 Introduction

Technology acquisition is a key process in managing technology [1]. Acquisition activities are concerned with decisions about the way to effectively incorporate selected technologies within organisations. Collaboration is an option to acquire technology, in particular when firms have a strong familiarity with a particular market or product but are unfamiliar with the technologies or services embedded into the product [2].

V. G. Ortiz-Gallardo (✉) · D. Probert · R. Phaal
Centre for Technology Management, Institute for Manufacturing,
University of Cambridge, Cambridge, UK
e-mail: vgo20@cam.ac.uk

D. Cetindamar et al. (eds.), *Strategic Planning Decisions in the High Tech Industry*, DOI: 10.1007/978-1-4471-4887-6_8,

The importance of collaboration to acquire technology is widely acknowledged in the field of technology management; nevertheless, technology acquisition by collaboration is a topic that still requires attention. Collaboration can be defined as "...any activity where two or more partners contribute differential [or complementary] resources and know-how to agreed complementary aims." [3], p. 13. However, collaboration itself is a challenging activity. Scholars and managers have reported that collaborations do not always meet the original expectations; for instance, some studies find that more than half of technology collaborations fails [4].

A survey carried out by the Product Development Management Association (PDMA) revealed that more than 50 % of new-to-the-world product developments projects are carried out in collaboration [5]. Thus, given the fact that new-to-the-world products usually incorporate new technology, it is evident that technology acquisition through collaboration is a relevant issue in business. Considering these two figures, the frequency of collaborations in new product development (NPD) and the low success rate of technology collaborations, it is clear that there is a need for understanding the conditions that allow effective acquisition of technology by means of collaboration.

Scholars have explored the influence of different factors on the performance of technology collaborations. Factors that have been explored include the characteristics of the acquiring firm [6], project attributes [7], type of partner [8], inter-organisational interactions [6], technology characteristics [8] and competitive environment [9].

Research on the performance of technology collaborations is fragmented and published across the different literature streams. For example, literature on strategic and technology partnerships provides insights to explain why certain collaborations achieve success and why others fail. But this literature seems to underestimate the effect of technology uncertainty on the outcomes of partnerships. Thus, despite the volume of research on this topic, there is still a need for a holistic framework to understand all the different factors that may have influence on the performance and outcomes of technology collaborations. This chapter aims to close this gap.

The conceptual framework depicted in this chapter shows the activities and types of factors that may have influence on the acquisition of technology through collaboration reported in the academic literature. Activities in the acquisition process were drawn from technology acquisition processes proposed by different authors and factors were identified through an analysis of papers published between 1990 and 2011.

## 8.2 Technology Acquisition

Firms may have the capabilities and resources to internally develop the required technology to support strategic or tactic business goals, or they may opt to acquire it from an external source [10]. The development of a product or process embedding external technology finishes at the point where its performance is demonstrated in the final operative system under real conditions [11]. Thus, acquisition of technology from an external source usually implies further development work to implement the technology into a final product or process.

Acquisition of external technology comprises a series of activities and decisions, ranging from the definition of technology needs to the implementation of the technology into the recipient product or process. The process to acquire technology can be broadly summarised as a six-stage process, based on a review of published approaches summarised in Fig. 8.1 for comparison.

1. Definition of technology requirements.
2. Identification of available technology.
3. Technology evaluation and selection of the source.
4. Negotiation.
5. Implementation.
6. Value added audit and relationship management.

The activities associated with each of the stages are described below.

*Definition of technology requirements*. This stage comprises the initial activities of the acquisition. Activities in this stage aim to analyse the current situation of the acquiring firm and to identify the technology gaps to accomplish the business objectives. As a result of these activities, a technology acquisition plan may be elaborated.

*Identification of available technology.* The next group of activities aims to explore different technology alternatives that can address the needs of the firm. These activities may include the identification and classification of technology solutions, as well as scouting of potential technology providers. The result of such activities is an important input to outline a business case.

*Technology evaluation and source selection.* The following set of activities intends to evaluate the different technology alternatives in terms of their capability to address the need of the acquiring firm, and in terms of the source. This group of activities may include the demonstration of the technology to perform a particular application, and the assessment of the different acquisition sources and methods.

*Negotiation.* The negotiation phase includes activities that are related to the definition of the terms of the acquisition and agreement formalisation, such as confirmation of the business case, preparation of the contractual agreement and negotiation of the scope of the transaction with the providing firm.

*Implementation.* The acquisition is done with the implementation of the technology into a final product or process. The activities considered in this group are associated with transferring the technology to the final operative environment and solving operative issues to achieve the expected performance.

*Value added audit and relationship management.* After the technology has been implemented, the process concludes with an evaluation of the success of the acquisition. The last stage of the process comprises an assessment of both the value added by the acquisition and whether is possible or not to keep a long-term collaboration with the provider.

**Overall stages**

| | Definition of technology requirements | Identification of available technology | Technology evaluation and selection of the source | Negotiation | Implementation | Value added audit and relationship management |
|---|---|---|---|---|---|---|
| **Durrani et al. 1999** | Establish market place requirements | Identify technology solutions; Classify the technology solutions | Establish sources of technology acquisition | Finalise the technology acquisition decision | | |
| **Eskelin 2001** | Initiation; Planning | Research | Evaluation | Negotiation | Implemen-tation; Operation | |
| **Baines 2004** | Technology profiling; Establishing the requirements of technology | Find a technological solution; Form outline business case | Choose technology source; Demonstrate technology | Confirm business case | Implement technology | Post-investment audit |
| **Daim and Kocauglu 2008** | Technology gap analysis or problem identification | Identification of technology alternatives | Evaluation of the technology alternatives | Acquisition of the technology | | Impacts through the acquisition of technologies |
| **Cetindamar et al. 2010** | Goal setting | Finding technology suppliers | Choosing acquisition method | Contract preparation and negotiation | Technology transfer | Managing long-term collaboration |

**Fig. 8.1** Key activities in the acquisition of external technologies. Durrani et al. [30], Eskelin [31], Baines [32], Daim and Kocaoglu [33], Cetindamar et al. [34]

This section has highlighted the different activities that are key to the acquisition of external technologies. However, acquiring technology is not a straightforward task. There are a number of factors that may have influence on the results of the acquisition process. The next section will introduce the type of factors that may affect the acquisition of technologies that have been reported in the existing literature.

## 8.3 Type of Factors Affecting Acquisition of Technology by Collaboration

There are a number of factors that may influence the acquisition of technology when firms work in collaboration. Table 8.1 summarises the different types of factors that may have influence on the acquisition of technology by means of collaboration.

The types of factors affecting the acquisition of technology by collaboration have been reported in three streams of the academic literature:

1. Management of technology acquisitions,
2. Management of strategic and technology partnerships and
3. Technology integration in NPD.

Factors that can affect the outcome of technology collaborations seem to be related to the interaction between partners and to the specific circumstances of the acquisition project.

**Table 8.1** Types of factors affecting acquisition of technology

| Type of factors affecting technology acquisitions | Stream of literature | Supporting references |
|---|---|---|
| Factors affecting acquisition of technology can be divided as *endogenous* and *exogenous* | Management of technology acquisitions | [12–17] |
| Inter-organsational interactions are influenced by four categories of factors: *Strategic alignment, structural match, coordination capability*, and *contextual issues* | Management of strategic and technology partnerships | [8, 18–25] |
| Technology transfer success occurs through the optimal match between *technology uncertainty* and inter-organsational interaction | External technology integration in new product development | [8, 10, 26–29] |

### 8.3.1 Management of Technology Acquisitions

Academic papers addressing management of technology acquisitions suggest that there are a number of factors that may influence the success of technology acquisitions. Factors that are covered in this literature include: in-house R&D capabilities, bargaining power [12], urgency, technical complexity, collaboration method [13] technology assessment methods [14], availability of financial resources, strategic priorities, market risks [15], ease of access to technology intelligence reports, additional cost of utilising the technology, lack of understanding of the technology [16], industry competition basis and pace and magnitude of technological change [17].

Factors covered in this stream of literature can be divided into two categories: endogenous and exogenous. Endogenous factors can be managed during the acquisition project. This kind of factor includes issues such as means to identify technology options, skills to valuate and evaluate technology options, resources to acquire external technology, absorptive capacity and management capabilities. Exogenous factors, on the other hand, comprise factors related to the characteristics and settings of the project. This group includes issues such as competitive environment, market risks and technology characteristics.

Despite the fact that authors in this stream of the literature have pointed out a large number of factors, they have not given full attention to the particular factors that affect the acquisition process by means of collaboration.

### 8.3.2 Management of Strategic and Technology Partnerships

Issues related to technology collaborations can be also found in the literature addressing management of strategic and technology partnerships (see for example [18–24]). Research in this area focuses on inter-organisational factors that ultimately influence the effectiveness of technology transactions, in particular when products and technologies are developed in collaboration.

The fundamental assumption in this stream of the literature is that two or more organisations work together to achieve complementary aims. In these relationships two roles can be identified: an entity that receives or acquires technology or know-how and an entity that provides technology or know-how. This strand of the literature suggests that the success of technology collaborations depend upon four themes: *structural match, strategic alignment of the partners*, *coordination capability* and *contextual issues.*

*Structural match* is a condition for any collaborative project [20, 21, 24]. In collaborations, each partner brings a piece of key technology or know-how that is critical to the project. In bilateral collaborations for example, one of the partners often provides technical skills and knowledge and the other provides market knowledge or specifications of the final product.

*Strategic alignment* comprises factors that motivate partners to set up a collaboration agreement. Some aspects such as harmony of business plans, joint specification of appropriate configuration and common time frame for achieving goals contribute to achieving strategic alignment between partners [18, 22, 24].

*Coordination capability* includes factors that influence the quality of the inter-organisational interaction. Cultural differences are often barriers to coordinate collaborative work [18–20]. People from the partnering firms interact over the duration of the acquisition process; consequently, the quality of the coordination between parties influences the outcome.

Finally, *contextual issues* comprise the settings and circumstances that may influence the interaction between partners. Each partnership is unique and consequently is characterised by a unique combination of settings and circumstances. The competitive environment [24], partner's size [19, 25] and nationality of the partners [25] are factors that have influence on the performance of the partnership.

Literature addressing management of strategic and technology partnerships provides a structured approach to explain the performance of collaborations. But since this literature is not fully concerned with the effective transference of technology between partners, it gives very little attention to the potential influence of the nature of the technology on the collaboration outcomes.

### 8.3.3 Technology Integration in NPD

The third strand of the literature comprises a series of publications that have analysed the influence of the characteristics of the technology on the effective integration of technology in the collaborative development of new products (see [8, 10, 26–28]). Technology characteristics such as complexity, tacitness, novelty, and maturity have been explored in this stream of the literature. Hacklin et al. [29] argue that the nature of a technology determines the success or failure of a development project, no matter how well it can be managed. Ettlie and Pavlou [28] highlight the importance of technology uncertainty in the outcomes of collaboration projects. They point out that technology transfer success occurs through the optimal match between technology uncertainty and inter-organisational interaction. Cui et al. [8] explore the success drivers in collaborations related to the maturity of the technology. They observe that successful acquisition of embryonic technologies is characterised by trust and communication, organisational stability of the partner and defined goals. In the case of acquisition of mature technologies, they found that project success depends also on compatibility of the technology with the existing systems architecture and flexibility of the partner to adjust components to accommodate broader systems requirements.

The three streams of the literature discussed above provide different perspectives on the types of factors that may influence the effective acquisition of technology by collaboration. Literature on technology acquisition provides a large number of factors that may influence the effective acquisition of technology, but it

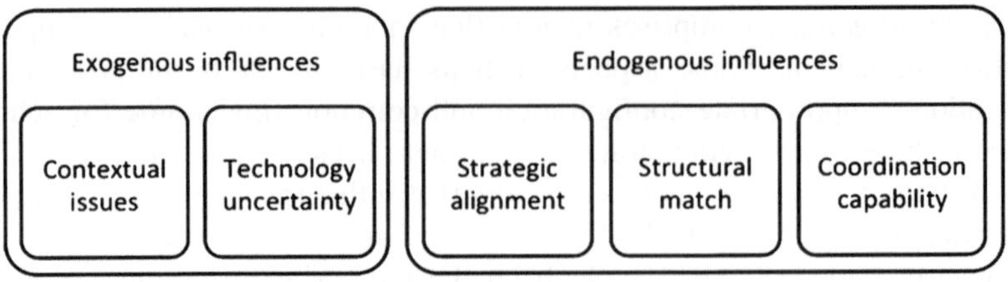


**Fig. 8.2** Type of factors that affect acquisition of technology by means of collaboration

gives little attention to inter-organisational issues that may affect the collaborative work. Literature on management of strategic and technology partnerships provides a wider perspective on the type of factors that have influence on the performance of collaborations; however, this literature does not give attention to the factors that may have influence on the effective transference of technology between partners. Finally, literature on technology integration in NPD recognise that both inter-organisational issues and technology characteristics have influence on the effective transference of technology between partners; nevertheless, compared to the two streams of the literature described before, the number of publications on technology integration in NPD is still low. Most of the literature in this stream has been focused on exploring the key success drivers to transfer effectively technology in collaborative NPD.

The three streams of the literature explored in this section provide complementary types of factors that may affect the effective acquisition of technology by means of collaboration. The analysis of papers in these streams of the literature suggests that five types of factors may have impact on the outcomes of technology collaborations: *technology uncertainty*, *contextual issues*, *strategic alignment*, *structural match* and *coordination capability*. These five types of factors can be further divided as exogenous and endogenous (Fig. 8.2). Exogenous influences include *contextual issues* and *technology uncertainty*. These factors define the settings of the acquisition project. Firms may not be able to control those factors over the acquisition process. Endogenous influences, on the other hand, comprise factors that emerge as consequence of the interaction between the acquiring and providing firm. These factors are separated into in three types: *strategic alignment*, *structural match* and *coordination capability*.

## 8.4 Factors Affecting Acquisition of Technology by Collaboration

The five types of factors presented in the previous section provide a layout to categorise the factors that have been explored in previous research. A review of papers published between 1990 and 2011 lead to the identification of 21 factors that have influence on technology collaborations. Table 8.2 shows a summary of

**Table 8.2** Factors affecting acquisition of technology by means of collaboration explored in previous research

| Type of factor | Factor explored | Reference |
|---|---|---|
| Contextual issues | Acquiring firm characteristics | [5, 6, 10, 35–39] |
| | Project characteristics | [6, 7, 21, 25, 26, 40, 41] |
| | Competitive environment | [6, 9, 40, 42–44] |
| | Type of partner | [8, 20, 45–48] |
| Technology uncertainty | Technology dynamism of industry | [9, 40] |
| | Technology characteristics (novelty, maturity and complexity) | [8, 10, 26–28, 37, 49, 50] |
| Strategic alignment | Incentive alignment | [38, 51] |
| | Motivations | [52–55] |
| | Importance of the project | [10, 21, 52, 56, 57] |
| | Partners' growth strategy | [58–60] |
| | Management of IPR | [51] |
| Structural match | Resources in exchange | [36] |
| | Partner identification and selection | [54, 56, 61–63] |
| | Partner competence | [38] |
| Coordination capability | Development team's effectiveness | [54, 64, 65] |
| | Trust | [21, 38, 52, 56] |
| | Project management | [52, 66, 67] |
| | Partners' experience in collaborations | [41, 68] |
| | Management of collaborations | [4, 18, 22, 46, 69–71] |
| | Inter-organisational interaction | [6, 10, 49, 56, 72, 73] |
| | Clear goals | [38, 54] |

the factors explored by several authors. Some factors seem to be widely explored; for example, acquiring firm characteristics, project characteristics, competitive environment, type of partner, technology characteristics, importance of the project, partner identification and selection, experience in collaborations as well as inter-organisational interaction. The factors comprised within each of the five types of factors are described below.

## 8.4.1 Contextual Issues

As previously mentioned, contextual issues cover situations linked to external events or settings that eventually may change, but they are not under the control of the partnering firms. This type of factor includes acquiring firm characteristics, project characteristics competitive environment and type of partner. These factors may have direct influence on the relationship and ultimately on the effective implementation of the technology.

### 8.4.2 Technology Uncertainty

Technology uncertainty comprises the characteristics of the technology and the technology dynamism of the industry. Technology characteristics such as novelty, maturity and complexity have been recognised as important aspects to consider in the effective implementation of technologies. These characteristics influence the level of risk, time, resources and amount of development work required to bring the technology to the point where it can be implemented into a product or process. The technology dynamism of the industry relates to the frequency of technology change.

### 8.4.3 Strategic Alignment

Academic literature points out that strategic alignment is crucial to set-up a collaborative agreement. Strategic alignment comprises factors that influence the establishment of a common vision and understanding of the benefits to each participant. Incentive alignment, motivations, importance of the project, partner's growth strategy and management of resulting intellectual property are identified as influential factors to set-up collaborations.

### 8.4.4 Structural Match

Structural match is recognised as a driver for establishing collaborative efforts and selecting a partner to work with. Organisations are likely to enter into a collaborative agreement when another entity possesses complementary resources. Structural match includes the following factors: resources in exchange, partner identification and selection as well as partner competence.

### 8.4.5 Coordination Capability

Coordination capability includes those factors that influence the quality of the communication between the partners over the duration of an acquisition project. Factors included in this group are the effectiveness of the development team, trust, project management, experience of the partners in collaborations, management of collaboration, inter-organisational interaction and clear goals.

These five types of factors can be combined with the stages of technology acquisition into a framework that helps to understand and manage the complexity of technology collaborations, as described in the next section.

## 8.5 A Conceptual Framework

Based on the analysis of the existing literature, a conceptual framework is presented. The framework points out factors and activities that influence the acquisition of technology by collaboration (Fig. 8.3). The framework considers two dimensions: one longitudinal and the other transactional. The longitudinal dimension comprises the groups of activities that are carried out to acquire external technologies. On the other hand, the transactional dimension considers the types of factors that may influence the outcomes of the acquisition by means of collaboration. The importance of the factors may vary at every stage of the acquisition process.

The framework does not provide insights on possible interrelations between factors and activities, but it offers a perspective of factors and activities that may have a major importance on the acquisition of technology by collaboration. The framework highlights factors that should be taken into consideration in managing technology collaborations. Researchers and practitioners can use it in further research as a reference to explore the importance of each type of factor under different circumstances.

The framework has been subject to testing and refinement through industrial consultation. A number of interviews with practitioners and seven case studies

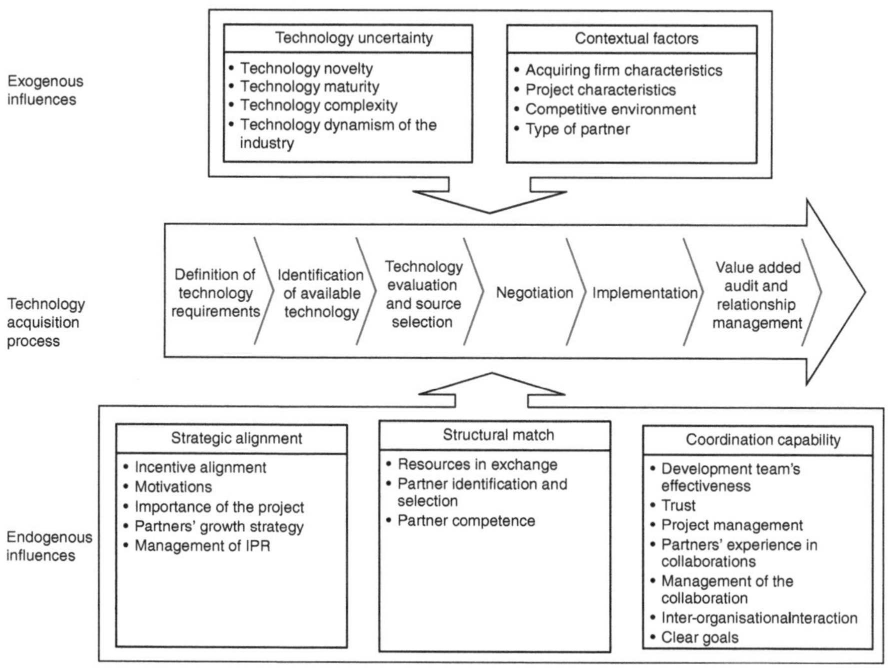


**Fig. 8.3** Activities and factors affecting the acquisition of technologies by means of collaboration

**Table 8.3** Characteristics of case studies that have provided support to the conceptual framework

| Case No. | Outcome target industry | Technology domain | Project objective |
|---|---|---|---|
| Case 1 | Biofuels | Manufacturing technology | Developing a commercial process to produce cellulosic ethanol |
| Case 2 | Biofuels | Manufacturing technology | Developing a process to produce biobutanol |
| Case 3 | Oil and gas production | Information processing technology | Developing a tool for downhole monitoring |
| Case 4 | Gas transportation and storage | Information processing technology | Developing a system for monitoring the integrity of gas pipelines |
| Case 5 | Petrochemicals | Manufacturing technology | Increasing the range of polyethylene grades |
| Case 6 | Construction industry | Product technology | Scaling up the production of an insulating material for applications in the construction industry |
| Case 7 | Paint and coatings | Product technology | Developing a new resin system for high performance coatings applications |

supported both the acquisition process and the types of factors that affect the outcomes of technology collaborations. Table 8.3 summarises the characteristics of seven case studies that have provided support to the conceptual framework.

## 8.6 Concluding Remarks

Collaboration is a frequent activity in industry. Companies collaborate with suppliers, customers, competitors, R&D centres and universities because of different reasons. Nevertheless, collaborations are not always managed in a systematic way. Frequently, companies choose to work with a particular partner because of successful previous projects. Often only a few people within firms are able to manage the relationships with partnering firms effectively.

In technology planning activities, firms may identify a number of technologies that should be acquired. Such technologies can be developed internally, acquired from an external source or developed in collaboration with a partner. However, only those firms who have worked intensively with third parties may be confident about relying on partnerships as a means to acquire technology, while those firms that are not familiar with partnerships may consider technology acquisition by collaboration as a potential risk.

The framework for acquiring technology by means of collaboration is presented in this chapter. The framework has been built through the analysis of the existing literature. It offers a wide perspective of the influential factors that lead to an effective acquisition of external technologies by collaboration. Previous literature has explored the influence of a number of factors upon the outcomes of technology acquisitions, technology partnerships and the integration of external technologies. However, few scholars have attempted to provide an integrated overview of the influential factors to achieve the effective implementation of external technology by means of collaboration (see for example [20–22]).

The conceptual framework depicted in this chapter compiles a series of influential factors that may affect the effective acquisition of technology through collaboration. The five types of factors identified in the framework capture the different issues that have been explored in previous research. The framework provides a guide to understand the factors that influence the performance of technology collaborations aimed to acquire and implement a technology into a new product or process.

Managing collaborations is challenging, and most managers develop these skills by means of practice. It is difficult to collaborate with a partner for the first time, and usually, the first issue that firms have to address is the selection of an appropriate partner to work with. The ability to effectively manage collaborations is fundamental for those firms that need to access key resources possessed by other companies.

This chapter summarises a number of factors that may influence the effective acquisition of technology by collaboration. Factors can be divided into five types: contextual issues, technology uncertainty, strategic alignment, structural match and coordination capability. The conceptual framework makes evident that managing collaborations is not a trivial task. Managers may have to deal not only with technology uncertainty but also with other business issues. Thus, in technology planning activities, managers have to take into consideration different factors when deciding the technologies that will be acquired through collaboration. The framework presented in this chapter recognises not only the types of factors that may affect the performance of the collaboration, but also the key stages in the acquisition process. Interviews with practitioners and the analysis of seven case studies provide support to the conceptual framework.

## References

1. Gregory M (1995) Technology management: a process approach. Proceedings of the institution of mechanical engineers. Part B: J Eng Manuf 209:347–356
2. Roberts EB, Berry CA (1985) Entering new businesses : selecting strategies for success. Sloan Manage Rev 26:3–17
3. Dogson M (1993) Technology collaboration in industry. Routledge, London
4. Duysters G, Kok G, Vaandrager M (1999) Crafting successful strategic technology-partnerships. R&D Manage 29:405–420

5. Barczak G, Griffin A, Kahn KB (2009) PERSPECTIVE: trends and drivers of success in NPD practices: results of the 2003 PDMA best practices study. J Prod Innov Manage 26:3–23
6. Buse S, Armonaitis P (2010). Outsourcing of R&D: chances and risks. In: 2010 IEEE International conference on management of innovation and technology (ICMIT), Singapore, pp 2–5
7. Bhaskaran SR, Krishnan V (2009) Effort, revenue, and cost sharing mechanisms for collaborative new product development. Manage Sci 55:1152–1169
8. Cui Z, Loch C, Grossmann B, He R (2012) How provider selection and management contribute to successful innovation outsourcing: an empirical study at siemens. Prod Oper Manage 21:29–48
9. Ang SH (2008) Competitive intensity and collaboration: impact on firm growth across technological environments. Strateg Manag J 29:1057–1075
10. Stock GN, Tatikonda MV (2008) The joint influence of technology uncertainty and interorganizational interaction on external technology integration success. J Oper Manage 26:65–80
11. Steele LW (1989) Managing technology: the strategic view. McGraw-Hill, New York
12. Sen F, Rubenstein AH (1990) An exploration of factors affecting the integration of in-house R&D with external technology acquisition strategies of a firm. IEEE Trans Eng Manage 37:246–258
13. Steensma HK (1996) Acquiring technological competencies through inter-organizational collaboration: an organizational learning perspective. J Eng Tech Manage 12:267–286
14. Henriksen ADP (1997) A technology assessment primer for management of technology. Int J Technol Manage 13:615–638
15. Durrani TS, Forbes SM, Broadfoot C (1999) An integrated approach to technology acquisition management. Int J Technol Manage 17:597–617
16. Slowinski G, Stanton SA, Tao JC et al (2000) Acquiring external technology. Res Technol Manage 43:29–35
17. Ranft AL, Lord MD, Carolina N (2002) Acquiring new technologies and capabilities: a grounded model of acquisition implementation. Organ Sci 13:420–441
18. Bronder C, Pritzl R (1992) Developing strategic alliances: a conceptual framework for successtil co-operation. Eur Manage J 10:412–421
19. Williams RG, Lilley MM (1993) Partner selection for joint-venture agreements. Int J Project Manage 11:233–237
20. Marxt C, Link P (2002) Success factors for cooperative ventures in innovation and production systems. Int J Prod Econ 77:219–229
21. Kim Y, Lee K (2003) Technological collaboration in the Korean electronic parts industry: patterns and key success factors. R&D Management 33:59–76
22. Emden Z, Calantone RJ, Droge C (2006) Collaborating for new product development: selecting the partner with maximum potential to create value. J Prod Innov Manage 23:330–341
23. Ettlie JE, Pavlou PA (2006) Technology-based new product development partnerships. Decis Sci 37:117–147
24. Hacklin F, Marxt C, Fahrni F (2006) Strategic venture partner selection for collaborative innovation in production systems: a decision support system-based approach. Int J Prod Econ 104:100–112
25. Paixao-Garcez M, Sbragia R, Kruglianskas I (2010) The selection of partners in non-equity bilateral alliances: some qualitative evidences from the Brazilian Petrochemical leader. PICMET, Thailand
26. Tatikonda MV, Rosenthal SR (2000) Technology novelty, project complexity, and product development project execution success: a deeper look at task uncertainty in product innovation. IEEE Trans Eng Manage 47:74–87
27. Tatikonda MV, Stock GN (2003) Product technology transfer in the upstream supply chain. J Prod Innov Manage 20(6):444–467
28. Stock GN, Tatikonda MV (2004) External technology integration in product and process development. Int J Oper Prod Manage 24:642–665

29. Souder WE (1987) Managing new product innovations. Lexington Books, Lexington
30. Durrani TS, Forbes SM, Broadfoot C (1999) An integrated approach to technology acquisition management. Int J Technol Manage 17:597
31. Eskelin A (2001) Technology acquisition: Buying the future for your business. Addison Wesley, New Jersey, USA
32. Baines T (2004) An integrated process for forming manufacturing technology acquisition decisions. Int J Oper Prod Manage 24:447–467. doi:10.1108/01443570410532533
33. Daim TU, Kocaoglu DF (2008) How do engineering managers evaluate technologies for acquisition? a review of the electronics industry. Eng Manage J 20:44–52
34. Cetindamar D, Phaal R, Probert D (2010) Technology management: activities and tools. Palgrave MacMillan, Basingstoke
35. Robertson TS, Gatignon H (1998) Technology development mode: a transaction cost conceptualization. Strateg Manag J 19:515–531
36. Das TK, Teng B (2000) A resource-based theory of strategic alliances. J Manage 26:31–61
37. Galbraith CS, Ehrlich SB, DeNoble AF (2006) Predicting technology success: identifying key predictors and assessing expert evaluation for advanced technologies. J Technol Transfer 31:673–684
38. Cui Z, Loch C, Grossmann B, He R (2009) Outsourcing innovation. Research technology management 52(6):54–63
39. Minshall T, Mortara L, Valli R, Probert D (2010) Making asymmetric partnerships work. Res Technol Manage 53:53–63
40. Lambe CJ, Spekman RE (1997) Alliances, external technology acquisition and discontinuous technological change. J Prod Innov Manage 14:102–116
41. McCutchen WW Jr, Swamidass PM, Teng B-S (2008) Strategic alliance termination and performance: the role of task complexity, nationality, and experience. J High Technol Manage Res 18:191–202
42. Narula R, Hagedoorn J (1999) Innovating through strategic alliances: moving towards international partnerships and contractual agreements. Technovation 19:283–294
43. Vilkamo T, Keil T (2003) Strategic technology partnering in high-velocity environments—lessons from a case study. Technovation 23:193–204
44. Yasuda H (2005) Formation of strategic alliances in high-technology industries: comparative study of the resource-based theory and the transaction-cost theory. Technovation 25:763–770
45. Kaufman A, Wood CH, Theyel G (2000) Collaboration and technology linkages: a strategic supplier typology. Strateg Manag J 21:649–663
46. Miotti L, Sachwald F (2003) Co-operative R&D: why and with whom? an integrated framework of analysis. Res Policy 32:1481–1499
47. Belderbos R, Carree M, Lokshin B (2004) Cooperative R&D and firm performance. Res Policy 33:1477–1492
48. Knudsen MP (2007) The relative importance of interfirm relationships and knowledge transfer for new product development success. J Prod Innov Manage 24:117–138
49. Steensma HK, Corley KG (2000) On the performance of technology-sourcing partnerships: the interaction between partner interdependence and technology attributes. Acad Manag J 43:1045–1067
50. Ragatz GL, Handfield RB, Petersen KJ (2002) Benefits associated with supplier integration into new product development under conditions of technology uncertainty. J Bus Res 55:389–400
51. Mehlman SK, Uribe-saucedo S, Taylor RP et al (2010) Better practices for managing intellectual assets in collaborations. Res Technol Manage 53(1):55–66
52. Littler D, Leverick F, Bruce M (1995) Factors affecting the process of collaborative product development: a study of UK manufacturers of information and communications technology products. J Prod Innov Manage 12:16–32
53. Chiesa V, Manzini R (1998) Organizing for technological collaborations: a managerial perspective. R&D Manage 28:199–212

54. Petersen KJ, Handfield RB, Ragatz GL (2005) Supplier integration into new product development: coordinating product, process and supply chain design. J Oper Manage 23: 371–388
55. Ford S, Probert D (2010) Why do firms acquire external technologies? understanding the motivations for technology acquisitions. Technology management for global economic growth (PICMET). Phuket, pp 1–4
56. Wagner SM, Hoegl M (2006) Involving suppliers in product development: insights from R&D directors and project managers. Ind Mark Manage 35:936–943
57. Lunnan R, Haugland SA (2008) Predicting and measuring alliance performance: a multidimensional analysis. Strateg Manag J 29:545–556
58. Pavia TM (1990) Product growth strategies in young high-technology firms. J Prod Innov Manage 7:297–309
59. Cardozo R, McLaughlin K, Harmon B et al (1993) Product-market choices and growth of new businesses. J Prod Innov Manage 10:331–340
60. Yasuda H, Iijima J (2005) Linkage between strategic alliances and firm's business strategy: the case of semiconductor industry. Technovation 25:513–521
61. Das TK, Teng B (2003) Partner analysis and alliance performance. Scand J Manag 19:279–308
62. Witzeman S, Slowinski G, Dirkx R et al (2006) Harnessing external technology for innovation. Res Technol Manage 49:19–27
63. Lee S et al. (2010) Strategic partner selection for collaborative R&D: Literature-based technology intellignece. In: R&D management conference, Manchester, UK, 30 June–2 July 2010
64. Jassawalla AR, Sashittal HC (1998) An examination of collaboration in high-technology new product development processes. J Prod Innov Manage 15:237–254
65. Gerwin D (2004) Coordinating new product development in strategic alliances. Acad Manag Rev 29:241–257
66. Tidd J, Bodley K (2002) The influence of project novelty on the new product development process. R&D Manage 32:127–138
67. Fraser P, Farrukh C, Gregory M (2003) Managing product development collaborations—a process maturity approach. Proceedings of the institution of mechanical engineers. Part B: J Eng Manuf 217:1499–1519
68. Reuer JJ, Zollo M (2005) Termination outcomes of research alliances. Res Policy 34:101–115
69. Cagliano R, Chiesa V, Manzini R (2000) Differences and similarities in managing technological collaborations in research, development and manufacturing: a case study. J Eng Tech Manage 17:193–224
70. Chesbrough H, Schwartz K (2007) Innovating business models with co-development partnerships. Res Technol Manage 50(1):55–59
71. Slowinski G, Hummel E, Gupta A, Gilmont ER (2009) Effective practices for sourcing innovation. Res Technol Manage 52:27–35
72. McCutcheon D, Stuart FI (2000) Issues in the choice of supplier alliance partners. J Oper Manage 18:279–301
73. Fliess S, Becker U (2006) Supplier integration—controlling of co-development processes. Ind Mark Manage 35:28–44

# Chapter 9
# Decision Making for New Technology: A Multi-Actor, Multi-Objective Method

**Scott W. Cunningham and Telli E. van der Lei**

**Abstract** Technology managers increasingly face problems of group decision. The scale and complexity of research, development, and alliance efforts in emerging fields of technology mandate a correspondingly sophisticated form of group coordination. Choices made include the selection of projects, the choice of investment alternatives, and the formation of technology licensing agreements. Multi-criteria decision analysis (MCDA) methods are often used to help decision makers in such situations. This chapter explores an approach closely related to MCDA, known as exchange modeling. Exchange modeling incorporates actor preferences, and assumptions about the play of the game, to better examine the resulting preferences of groups. The advantage of this method is that the results provide an improved prescription for strategy, given the constraints of preferences and the existing alliance structures. The model is motivated based upon the needs of technology managers in new, converging fields of technology. The model is formally analyzed using operations research techniques.

Reprinted from Technological Forecasting and Social Change, 76 (1), Cunnigham and van der Lei, Decision making for new technology: A multi-actor, multi-objective method, 26–38, (2009), with permission from Elsevier.

S. W. Cunningham (✉) · T. E. van der Lei
Delft University of Technology, Delft, Netherlands
e-mail: S.Cunningham@tudelft.nl

D. Cetindamar et al. (eds.), *Strategic Planning Decisions in the High Tech Industry*, DOI: 10.1007/978-1-4471-4887-6_9,

## 9.1 Introduction

Central to the modern conception of decision making is the concept of value. Values, as codified by von Neumann and others, specifically entail the idea of value at risk [1]. Thus, concepts of decision making under risk, such as lotteries, are useful both for assess the value of a decision, and in assisting the analyst in prescribing an appropriate course of action [2]. Multi-criteria decision analysis (MCDA) uses the concept of value to prescribe desirable trade-offs between decision alternatives [3]. Game theory uses value to select an appropriate strategy faced with environmental or strategic uncertainty [4]. Thus, two streams of decision-making theory have arisen from the identical foundations of decision making and value.

For both streams of decision making, value is exogenous to the model. Values are at the discretion of the decision maker, and are not affected by interactions with the social or economic environment. A more endogenous perspective on value creation is offered by cooperative game theory. In these models, participants bargain or negotiate to create new sources of value [5, 6]. However, cooperative game theory suffers from a surplus of prescriptive recommendations. There are multiple competing prescriptive recommendations for fairness in negotiation. Prescriptions for negotiated outcomes may vary widely according to the specifics of the case [7, 8].

The concept of revealed preferences is one method for operationalizing the concept of decision-making value [9, 10]. This entails the measurement of actual decision-making behavior and reverse inference about the nature of decision-making values which are presumed to have guided the decision. Thus, as analysts we infer the values of a decision maker given evidence of prior decisions. Another leading technique is value elicitation [11, 12]. In this approach, the decision analyst queries the decision maker to reveal their preferences. Further guidance is provided by decision structuring, and the resulting consistency that decision analysis introduces to decision making. Decision makers, however, are constrained in their decision-making capability—not just because of their values, but also because of the strategic context within which decisions are being made. Preferences, as revealed by the decisions of real actors are guided by values but are also strongly shaped by strategy.

The goal of this chapter is to more thoroughly unify the literature on multi-criteria and multi-actor decision analyses. In particular, we relate an exchange modeling approach, which has long been used in economics, systems engineering, and sociology, to the domain of management of technology. The assumptions of exchange modeling are compared and contrasted with those used in MCDA. The resulting comparison lays bare the challenges of preference aggregation in a group setting. The exchange modeling approach cannot solve the matter of preference aggregation, but it can introduce additional factors which help decision-makers prepare for the way in which negotiations and exchanges are likely to occur in a multi-actor setting. Most importantly, the chapter presents a vision of quantitative decision-support given strategic modeling. This vision, which is consistent with the tradition of decision analysis, views consistent execution of preferences in a strategic setting an important prescriptive goal. The results of this research may help to achieve a more design-oriented approach to guiding management of technology decisions.

The structure of the chapter is as follows. Section 9.2 considers previous work in exchange modeling. Section 9.3 offers a brief survey of the literature on strategic alliances and technology management. The section further discusses the role of decision analysis in strategic and multi-criteria problem formulation. The section further presents a generic mathematical model of exchanges and alliances. This section makes explicit the behavioral assumptions of the model, which are encoded in a set of utility functions for actors. Section 9.4 discusses how the exchange model may be parameterized, drawing upon empirical results from decision analysis, economics, sociology, and public administration. Section 9.5 provides a specific example of the decision model in the context of a corporate alliance problem. Section 9.6 concludes the chapter with main conclusions and points for future research.

## 9.2 Previous Work

In the following section, we review the literature concerned with prescriptive models for strategic decision making. Such models have been used in both the public and private sectors. Coleman pioneered work in modeling social exchange mechanisms [13–15]. Coleman postulated that actors exchanged control over issues to achieve desirable outcomes. This literature argues that deliberative processes in politics operate as a form of exchange. This is a valuable basis to begin consideration of similar models in the management of technology domain. The work has been critiqued by subsequent researchers [16–18]. Perhaps the most significant critique of the work is that actors have limited access to the halls of power. Thus, the conditions for the free exchange of issues are not met in practice.

Thus, the primary finding of research in this literature is that the configuration of actors strongly shapes the achievable outcomes of any group negotiation. The practical results of this work have informed the study of policy networks [19–22]. In the following paragraphs, we further examine the management and strategy of the literature in order to better understand parallel development and application of models for decision-making under strategy.

Alliance activity between firms is on the rise. Research has focused on the diversity of alliance types with examples including R&D consortia, manufacturing and distribution alliances, and product bundling arrangements. Researchers have examined the locus of competition between firms, alliances, and even between competing constellations of alliances. A key motivator in the research has been a desire to explain differences in firm performance. The relational view of the firm, for instance, suggests that firm performance lies in the quantity and quality of network ties available to the firm. Alliance ties are seen as both conduits of information as well as sanctioning mechanisms.

The optimal design of alliance strategies has been a topic of debate. One literature argues that dense networks of strong, closed ties are maximally advantageous to the firm. This literature argues that such networks promote consensus, shared values, and reduce coordination and transaction costs. An alternative perspective emphasizes the role of loose, distant, and diverse ties. Alliances which bridge structural holes, and thus generate new knowledge and innovative ideas, are to be particularly valued from this perspective [23, 24]. The literature demonstrates the significance of alliance activity for high technology companies. The significance of alliance activity in a variety of industries including semiconductors, telecommunications, and biotechnology is addressed [25–28]. More broadly, the issue has been characterized as a challenge of managing value creation with networks under distributed and decentralized control [29, 30].

In the following section, we compare and contrast four models of decision making, resulting in a progressively more complex depiction of group decision making. The typical engagement structure and goals of each of the models are further detailed. We create our own systems model of strategic decision making in order to better compare the assumptions and advisory practice behind these four kinds of decision models.

## 9.3 Basic Elements of the Model

The model of MCDA contains a single actor, making a decision against nature. The resultant delivery of desired outcomes is subject to risk from nature. Many MCDA techniques structure the decision-making process by introducing a value function which prescribes desirable trade-offs between all achievable outcomes. Outcomes may have multiple attributes which may be used in their assessment through the value function. A weight on different outcomes or attributes is used to consider a range of rational preferences intrinsic to the decision maker (Fig. 9.1).

The role of the decision analyst in a multi-criteria analysis is to assist the decision maker in clarifying the nature of the risks, outcomes, and value functions at stake in a given decision. A structured interview or engagement is used to determine the correct decision weights needed to best reflect the preferences of a given decision maker [11].

One basic group MCDA technique makes only a minor modification to this basic scheme. A group of stakeholders, who are party to a decision, are consulted to determine their decision preferences. Both a weighting scheme for individual decision makers is used, as well as an aggregating scheme to translate individual preferences into a recommendation for group action. The technique is problematic for several reasons not the least of which is the interpersonal comparison of utility. Furthermore, the expression of weights is subject to political gaming as participants seek to drive the process to personally preferred outcomes [31]. For these

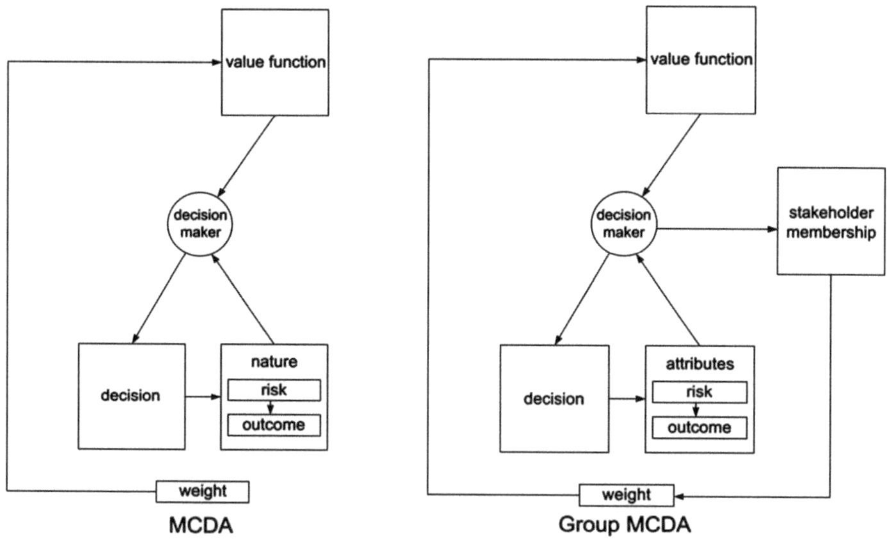


**Fig. 9.1** Model elements for MCDA and group MCDA

reasons, as well as others, a measure of decision consensus may be introduced by a facilitator. The group is thereby encouraged to identify, and resolve differences of opinion [32].

An alternative account of group decision processes is offered by the group exchange literature. This literature, inspired by computational models of the marketplace [33], suggests that even noneconomic forms of decision making may be described and governed by an exchange mechanism. Some additional elements to the model are required in order to embrace this exchange mechanism (Fig. 9.2).

In the exchange model decisions are made not against nature, but against self-interested players. A predetermined market structure dictates who may trade which outcomes. Individual decision weights are supplanted by the going market exchange rate. As a result, individuals with more say over outcomes of interest to the group have greater control over the decision-making process. Some minor changes in model terminology are required consistent with the exchange literature. Decision and nature become "control." The value function becomes "interest". Decision weights become the "exchange rate". Stakeholder members are governed by an "exchange structure" which regulates the nature of group decision making (Fig. 9.2).

Analysts operating with this model must identify stakeholders, exchange structures, as well as outcomes, and issues for a given case. Each stakeholder possesses a degree of control and interest over outcomes. The exchange rate between issues emerges from the model. The model may be used in engagement to assist group decision making by helping group members identify issues of interest, recognize points of compromise, and to drive outcomes to maximally satisfactory conclusions. Examples of decision aiding by means of the exchange model include

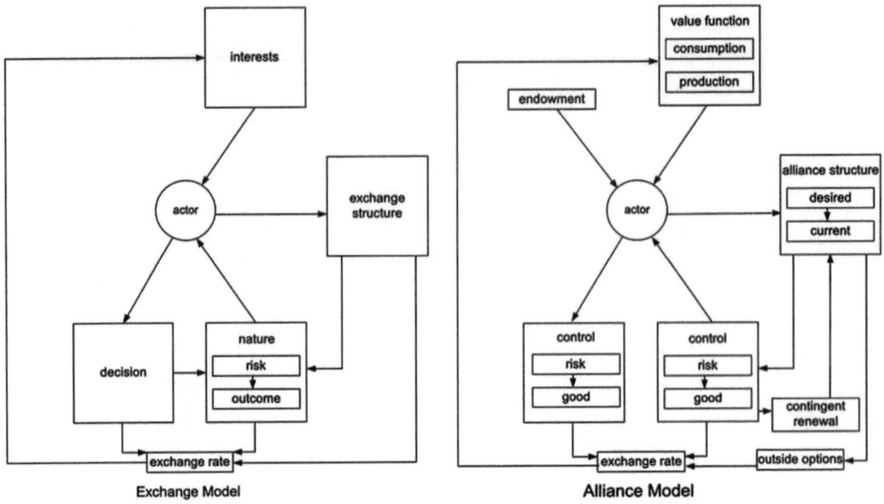


**Fig. 9.2** Basic model components in the exchange and alliances model

public sector example of land use negotiation [34], energy policy [18], as well as a private sector example involving the financial restructuring of a large farm cooperative [19].

The exchange model has been critiqued on a number of factors. Limited participant access may certainly change the equilibrium out come, as well as resultant prescriptions for action. Well-connected exchange members have a better bargaining position by falling back to "outside options" whenever necessary. This structural critique may be relatively limited since a more networked exchange may be readily incorporated into the basic model, as for instance, as incorporated into the works of Marsden [16, 17]. We will further consider the validity of these networked extensions to the exchange model in the conclusions of the chapter.

Another critique concerns the completeness of the representation of actors used in exchange models. Actor interests may be more complexly defined by considering both their range of desired outcomes ("consumption"), as well as their transformative capacity for action ("production"). Actors also may strongly differ in their base endowment of decision-relevant "goods". In many regards, the sociological exchange model abstracts some facets of economic life that may be usefully restored when modeling problems in the management of technology domain.

A still more serious basis for critique of exchange models is the idea that the fundamental exchange of control between actors entails a measure of risk. The management of risk and incentives between parties with incomplete or asymmetric information constitutes a significant body of ongoing research [35]. In this regard, the multi-criteria model provides a richer account of single- actor decision-making under risk than available in the conventional exchange model.

Furthermore, this risk may be mitigated by the exchange structure itself. Third-party recommendations between actors, and contingent renewal of agreements between actors, assist in sanctioning unreliable exchange behavior. A final enrichment to the model (Fig. 9.2) comes from the literature on strategic alliances. This literature examines the way in which alliance structures change over time, advantaging certain participants at the expense of others.

Analytical engagement using alliance models may take several forms. Modeling efforts may be intended as a theoretical explanation of real-world processes. The parameterization of the model to support decision making may therefore not be of primary concern. Alliance models may also be used for making policy recommendations. After fully specifying the model, an analyst may run a range of different policy experiments to determine likely changes in system structure and outcome. Work of this character historically has been done questions of a more economic than strategic character [33]. Policy experiments on alliance structure on the other hand remain a relatively new field of investigation.

Analytical work of this character no longer presumes unitary action (as for instance in much of the group MCDA literature). The analyst, therefore, forms prescriptions for action for a single actor, while taking into account a range of likely strategic behavior by other participants in the system. Value preferences for other actors in the system is not known, but is often imputed using techniques such as revealed preference. Exchange models lead to outcomes which describe the workings of a competitive economy, as well as situations of negotiation and bargaining. Thus, the model represents a good first cut at capturing the dynamics of many social games. As noted previously, the model originates in efforts to create a mathematical formulation for the theoretical concept of a free and competitive market [33]. Thus, the model has deep roots in neo-classical economics, and has been used extensively in the analysis of governmental interventions in the marketplace. Operations researchers have created mathematically related formulations in an effort to analyze network equilibrium [36]. Thus, the model has been used to explore equilibrium behaviors in both physical systems and actor systems.

These models are not linear optimization programs, but they are linear complementarity problems—optimization problems where there are distinct, yet multiply coupled objective functions. In this application, optimization is a representation of strategic decision making. As such, the optimization does not supplant the choices of the decision maker rather it enforces consistency between the choices of the decision maker and the known characteristics of the strategic environment. This is consistent with the long standing role of the decision analyst in helping a decision-maker frame a self-consistent set of preferences given an uncertain environment [2, 11, 12].

This tradition of creating quantitative models of exchange was further adopted by mathematical sociologists seeking theories to better describe political negotiation and bargaining. The model further diffused into the fields of public administration where it was used examined actor constellations seeking personal interest given public provision of goods. To our knowledge, the model has not yet been applied in a technology management context, but we believe the model has particular value in this setting given the group decision making or strategic context of management of technology problems.

The strategic management context often presents such issues of resource sharing or mutual investment in new technology in the context of negotiation or bargaining. Decision-makers begin with their desired outcomes and then attempt to formulate the rules of the game leading to these outcomes. This is an explicitly cooperative rather than a competitive framing of the problem. Some authors in strategic management explicitly adopt cooperative game theory concepts in their qualitative recommendations on business strategy [37]. Such strategically ill-formed problems are, on the face of it, very different from the economic setting where actors are "locked in" to a specific play of the game, given their preferences and budget. The question then becomes whether an explicitly competitive model can be appropriately used in a cooperative context. Models in the sociological tradition of exchange modeling are frequently used in a cooperative context, despite the ostensibly competitive character of the theory. As it turns out, many negotiation games—despite operating under very different actor dynamics—result in the same outcomes as that of a competitive economy. This matter was explored further in [8], formulating the exchange problem of a competitive economy as a matter of cooperation, and further characterizing the result using cooperative game theoretic concepts. Further investigation into exchange dynamics is needed to better match theory with applied problem solving.

## 9.4 Parameterizing the Model

In the following section, we discuss the problem of parameterizing the model. The problem is discussed in theory, examples are given in practice, and then a specific example from the management of technology domain is discussed. There are three quantities—control, interest, and weights. Any two of these quantities must be parameterized, and the third may be derived by the model. The model assumes the free exchange of resources by actors acting according to their own best interests. A fuller discussion of the dynamics of exchange follows this discussion.

### *9.4.1 Parameterizing Control and Interest*

A full range of techniques are available for parameterizing exchange models. Techniques common to decision analysis may be used. In addition, techniques drawing upon econometrics and strategic management may also be used. These techniques are discussed in further detail in the context of specific model parameters.

Control is an expression of an actor's values, and is directly analogous to the problem of scoring in MCDA. Explicit measures for eliciting scores are discussed in the MCDA literature. The same processes are useful for eliciting decision-maker interests. Belton and Stewart [11] provide three broad techniques in this regard: defining partial value functions, constructing qualitative value scales, and direct rating of the alternatives (pp. 121–134). Other authors in decision analysis also extensively cover assessment procedures for multi-attribute utility functions [12] (pp. 261–271). The elicitation procedure involves an interactive dialog between decision analyst and decision-maker. The decision-maker communicates preferences, and the decision analyst confirms structure and consistency of the values before encoding the preferences into a utility function.

Implicit methods are also used for evaluating interests and values. Samuelson pioneered the method of revealed preferences [9]. This method of economic valuation presumes that actors behave in a manner which consistent with their values. The analyst may induce preferences which are consistent with known observations of economic activity. Subsequently, revealed preferences have been used in wide ranging empirical studies including individual behavior, consumer and household choice, public choice for national services, and aggregated trade data [10]. The matter of assessing control is made somewhat easier by the fact that the model needs to appraise the sum total of control across all actors: thus control is made in comparison to other actors.

Interest expresses an actor's preferences for future outcomes. The matter is addressed in standard MCDA when, for instance, considering an actor's preferences over time [12] (pp. 473–513). Such utility functions, for instance, may be used to evaluate consumption and savings over the course of an individual's lifetime. Revealed preference methods here again play a role by assuming that preferences remain relatively constant over time. Economic specialization means that there is quite often a gap between what is desired and what is actually produced—in the part of consumers, the firm, and nations alike.

### *9.4.2 The Strategic Expression of Interest*

Future outcomes are also dependent upon the economic, political, and strategic choices of other actors in an economy. A classical decision analysis perspective envisages a decision maker as a synthesizer or amalgamator of the preferences of attendant stakeholders [12] (p. 516). More generally, however, there is a challenge in eliciting and synthesizing preferences in a strategic environment. The discipline of public choice, for instance, is much concerned with the strategic expression of preferences [31, 38]. In general, then it is methodologically more satisfactory to measure preferences indirectly, where strategic expression of preferences is unlikely to occur. This, however, requires a situation where actors have little bargaining power over another, as would occur for instance in a competitive marketplace. Direct expression of preferences may not be possible in a strategic context, and even where it is possible, the analyst must be aware of the possible strategic expression of expressed preferences. A final option presents itself and that lies in forming strategic hypotheses about the interests of other actors. Both public administration and strategic management rely in large part upon forming such strategic hypotheses about actor and stakeholder preferences. The quality of recommendations based upon such hypotheses depends both on an accurate appraisal of interests, as well as a rigorous analysis of the consequences of such preferences in a strategic setting. Given the considerable uncertainty about such preferences, the analyst must help evaluate a range of policies which are robust to strategic uncertainty. An example of decision support under systemic (although not necessarily strategic) uncertainty is provided by [39].

A final set of parameters in the model are the weights. These represent the real or anticipated trade-offs made by actors as they trade their control for greater interest in outcomes. If these values are real, then they may be measured directly using archival data. If anticipated, then they may either be calculated by the model, or elicited in consultation with decision makers. These exchange rates are direct analogs to weights from MCDA. The MCDA literature provides extensive guidance on the weight elicitation procedure with experts [12] (pp. 134–142).

### *9.4.3 Examples of Model Parameterization*

Timmermans used decision analysis techniques in concert with group conferencing to parameterize an exchange model [34]. Many authors in the field of exchange modeling use archival analysis [16–20] in their parameterization of models. Shoven and Whaley (in the closely related economic field of computational general equilibrium) are an example of an econometric approach to modeling national economies given exchange data [34]. Thoyer and co-authors created a model to support water resource negotiation. The authors supported decision making not through a comprehensive and exhaustively parameterized model, but rather through a robust exploration of alternative scenarios for negotiation [40].

Engelman provides an extensive account of using exchange models in a management of technology context [41]. TNO, the Dutch national laboratory, seeks guidance for formulating strategy in their semiconductor division. The author used decision analysis techniques to clarify the values of TNO, and expert opinion to parameterize the network structure of the industry, and the perceived strategic interest of competitors, customers, and alliance partners. The case suggests that gathering data in a management of technology context is challenging, but that the structuring of the data gathering process is assisted by an exchange analytic framework. On the other hand, the openness of TNO strategic networks, and the radical differences in economic scale between players in the network, presented a challenge to standard exchange models. These empirical cases demonstrate that a range of techniques for parameterizing exchange models have been used, techniques which include now traditional decision analysis techniques as well as more econometric methods.

## 9.5 Model Analysis and Application

The fundaments of the model are discussed in Appendix. At this point, it is most important to note from this derivation that actor interests, actor control, and the network structures permitted establish the going rate of exchange for goods within the system. In this section, we will provide a hypothetical example of the transactional approach for alliance formation in the microelectronics industry.

The case is inspired by a newsworthy supply chain failure [42]. The supply chain in question included notable participants such as Sony, Ericsson, and Nokia. The goal of the case is to demonstrate the mathematical modeling discussed in the previous section, as well as to provide a realistic motivating case for discussion. However, it is not the purpose of the case to provide specific strategic advice to the companies involved, and therefore we do not seek a careful parameterization of the network and utility structure of the participants involved.

Nonetheless, we offer for illustration purposes a list of possible participants in a microelectronics alliance (Table 9.1, Appendix). Figure 9.3 shows a hypothetical fully connected exchange between these 12 microelectronics suppliers and manufacturers. Together, the alliance participants are trading a number of different kinds of technical and business expertise for mutual benefit, including wireless technology, microelectronics manufacturing, and global marketing capability. Each alliance participant has its own native capability (control), as well as its own distinct desired outcomes (interests).

We might take for example alliance partner A, which has control over 10.4 % of the total market capability in developing new wireless technology. They are significantly less capable in microelectronics manufacturing and marketing (less than 1 % of the total in each). Their strategic goals are to capitalize on their wireless technology, and to expand their microelectronic manufacturing base. Direct marketing to consumers is not a priority. A corresponding specification of interests for this company might therefore be the following: (72 % wireless/26 % microelectronics/2 % marketing).

In the Coleman model exchange, participants are able to freely exchange goods to achieve their desired outcomes. Decision makers are subject to budgetary constraints, which are determined by the market valuation of the goods under their control. The calculations for this are described in Appendix. A market price for each commodity can be determined (Table 9.1). Then the equilibrium resource levels for each company can be determined (Table 9.2).

**Table 9.1** Equilibrium exchange rates for goods

| Good | Exchange rate |
|---|---|
| Wireless technology | 0.308 |
| Microelectronic manufacturing | 0.394 |
| Global marketing capability | 0.298 |

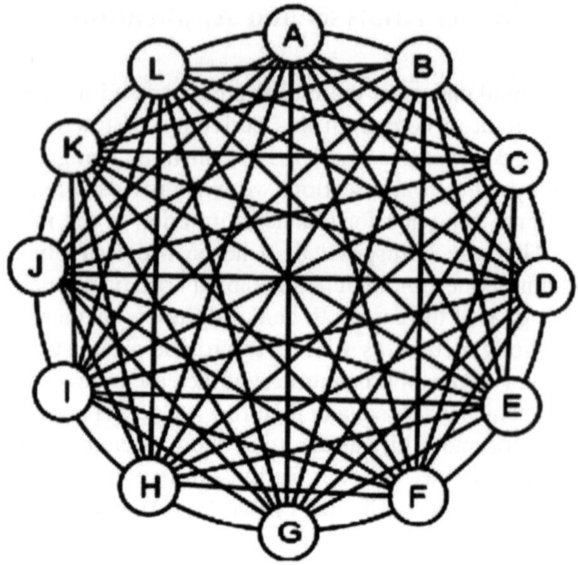


**Fig. 9.3** Full exchange structure

**Table 9.2** Equilibrium resource levels after exchange (matrix R)

| Firm | Market power |
|---|---|
| A | 0.077 |
| B | 0.055 |
| C | 0.069 |
| D | 0.124 |
| E | 0.022 |
| F | 0.136 |
| G | 0.179 |
| H | 0.026 |
| I | 0.067 |
| J | 0.083 |
| K | 0.082 |
| L | 0.081 |
| Total | 1.000 |

In this specific example, the market places a high premium on microelectronic manufacturing skills. Our representative company, company A, is highly capable in wireless technology but like many other companies in this hypothetical exchange members all place a high demand for microelectronic capabilities. As a result it fares comparatively poorly in the marketplace, receiving little of the good it needs to achieve its own strategic objectives.

Note that in this example all quantities (control, interest, resource levels, and prices) are unit less. The specific units and quantities are determined by the problem involved. We may drop units from the analysis without any loss of

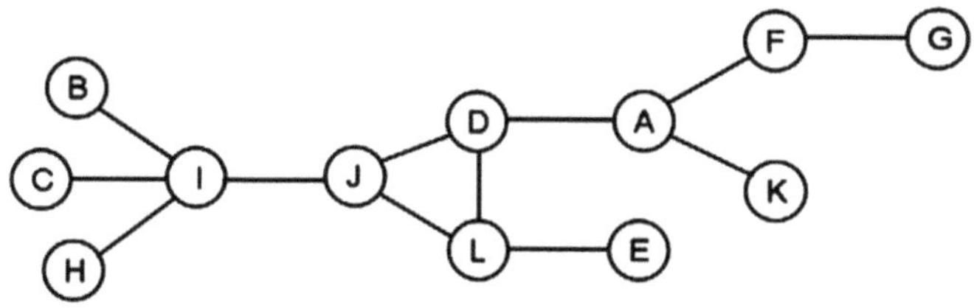


**Fig. 9.4** Network structure

generality. We may also select our units for prices and resource levels—the numeraire—as we wish. What is fundamental to the problem is the ratio of the price of goods and services, not the specific units of measurement [33].

We now expand the exchange model still further by hypothesizing a specific alliance structure which guides the permissible trades between participants (Fig. 9.4). Limited access between participants is the norm is such alliance activities, thus a networked model produces a greater fidelity of results to actual strategic behaviors.

In general, we would expect group outcomes to be lower (as free and open trading is no longer possible, and viable trades between participants are lost). However, some network participants, such as the centrally located companies J, D, and L may preferentially benefit from the exchange. As will be seen below, it is the specific configuration of actors and interests that dictates the network outcomes (Table 9.3).

In this example alliance, the configuration places a further demand on micro-electronic manufacturing, and wireless technology, already the scarce resources in the system. In contrast, the exchange rate for global marketing is further depressed. The networked exchange process further advantages certain actors compared with the free trade model of exchange (Table 9.4).

As can be seen our exemplar company, company A, is further disadvantaged by its position in the alliance constellation. Increasingly limited access to desired microelectronic skills further reduces the competitiveness of this company. In this specific example, most companies were advantaged by the network constellation and only three were made comparatively worse off (companies A, G, L).

In our last set of numerical experiments, we calculate the potential economic advantage or disadvantage of adding or subtracting alliance links in the network. We perform these experiments using company A as an example (Table 9.5).

**Table 9.3** Equilibrium exchange rates for goods under networked coordination

| Good | Exchange rate |
|---|---|
| Wireless technology | 0.347 |
| Microelectronic manufacturing | 0.403 |
| Global marketing capability | 0.250 |

**Table 9.4** Equilibrium resource levels after networked exchange (matrix R)

| Firm | Free exchange | Networked exchange |
|---|---|---|
| A | 0.077 | 0.040 |
| B | 0.055 | 0.086 |
| C | 0.069 | 0.090 |
| D | 0.124 | 0.096 |
| E | 0.022 | 0.084 |
| F | 0.136 | 0.116 |
| G | 0.179 | 0.039 |
| H | 0.026 | 0.108 |
| I | 0.067 | 0.148 |
| J | 0.083 | 0.120 |
| K | 0.082 | 0.032 |
| L | 0.081 | 0.041 |
| Total | 1.000 | 1.000 |

**Table 9.5** Value and marginal value of alliance partnerships

| Firm | Change in alliance structure | Equilibrium resources | Marginal value of exchange |
|---|---|---|---|
| A | Baseline | 0.0400 | 0.0000 |
| B | Add | 0.0394 | −0.0006 |
| C | Add | 0.0397 | −0.0003 |
| D | Remove | 0.0396 | −0.0004 |
| E | Add | 0.0392 | −0.0008 |
| F | Remove | 0.0396 | −0.0004 |
| G | Add | 0.0396 | −0.0004 |
| H | Add | 0.0393 | −0.0007 |
| I | Add | 0.0395 | −0.0005 |
| J | Add | 0.0393 | −0.0007 |
| K | Remove | 0.0401 | 0.0001 |
| L | Add | 0.0398 | −0.0002 |

Company A is in the position of both being disadvantaged by its network position, and also having very little strategic recourse to bilaterally improve its position. Most single link additions or subtractions to the alliance structure deplete the equilibrium re-sources of the company. Removing one link, however, the link to company K, increases the equilibrium resource levels of the company. It is possible that dropping this alliance enables the company to redouble its efforts with the other more profitable alliances in its portfolio.

We conclude with some comments on the character of advice provided by the exchange model. Results from the exchange model are markedly different from MCDA. Certainly, the standard multi-criteria model may be useful in this setting by assisting decision makers to set their strategic interests for the company. However, the emergent weighting of key goods in this industry is determined in significant part by the alliance structure of the model, and the collective and strategic action of other players. Likewise, the assessment of marginal value of alliance partnerships is determined in large part by the embedded nature of the firm within this alliance constellation. Given the strategic character of action in the sector, a single group-decision model, where pooled evaluations of necessary new industrial enhancements are made, may not be feasible.

## 9.6 Conclusions and Future Research

The previous sections provided mathematical details of an exchange model, as well as a simple numerical example illustrating a microelectronics supply chain. Existing results from models of political activity are extended to a problem of management of technology. The model is also used to demonstrate new results, in the form of sample strategic advice on alliance formation. In particular, the chapter examined the consequences of restricted access to network resources in determining desired outcomes for a firm in an alliance formation situation. The model, if parameterized to a specific problem setting, may be used to evaluate the existing alliance networks, design new networks, and select new alliance partners. Parallel efforts in the field of network games are describing the underlying logic of strategic action in networks, and thereby providing new insight into the growth and formation of business and technology networks.

Perhaps the most pressing concern in further expanding and exploring the model lies in the assumptions of price equilibrium in a network. The networked extensions to the exchange model, as proposed by Marsden [16, 17], argue that a networked alliance should be able to converge to a single, globally applicable exchange rate for goods and services on the network. Recent research suggests that these unitary valuations, in absence of external institutions such as government regulation, will not hold at equilibrium [43]. Individual companies can and will profit by raising prices locally on the network. The dual to this problem of local pricing is the alliance allocation decision: given scarce resources, where must alliance trades be made to maximally enhance profitability? Prior results in the field of cooperative games suggest that an equilibrium coalition structure requires maintaining a balanced investment by all alliance partners, where certain alliances are more likely to be favored partnerships [8].

Other results on network equilibrium are being generated within the field of network games, as well as within the transportation policy literature in the context of multi-commodity flows [44]. Coleman initiated efforts to research the role of trust within exchange networks. More recent efforts have examined the role of iterated games played on a network, such as the game "iterated prisoner's dilemma with choice and refusal". Information economics, and especially contract theory, are investigating the nature of strategic relationships when there is asymmetric information, such as about the reliability of a partner [35]. This area is also a fruitful for future investigation.

The model used in this chapter is a system of linear value functions. Linear functions are also used in many MCDA situations where more complex representations of values may be inappropriate given available information about preferences. Nonetheless, further research should examine the consequences of using alternative functional forms for value functions. Economists use functions which are behaviorally more complex, but also readily parameterized given records of economic activity [33].

The model presents a general framework for modeling free exchange and open negotiations. The model is consistent with a wide range of possible strategic processes. This is both an advantage and a limitation of the model. Further explorations should be under-taken with the model to capture some of the additional complexities of exchange, including information asymmetry, and bargaining power. This additional exploration would also help to falsify the conclusions of a widely applicable, yet quite generic, representation of actor dynamics.

This chapter has pursued a tighter integration between models of multi-criteria analysis and multi-actor modeling. The chapter created an exchange model of decision making, and related the resultant group valuation of outcomes to a broad class of multi-criteria decision-making problems. These two approaches share a common theory, and many of the same empirical techniques for capturing the interests and preferences of decision makers. However, further diffusion of exchange modeling has been hampered by the often overly technical and strictly disciplinary orientation of many studies. The resultant combination of modeling approaches offered new insights, extensions, and critiques to the field of group decision making. We hope the sample problem presented will stimulate new empirical applications in the field of technology management.

**Acknowledgments** The authors appreciate the comments of an anonymous reference whose comments helped improve the chapter. The authors also appreciate the empirical work of Engelman which enabled us to better formulate the problem of strategic modeling in the domain of management of technology. The chapter was originally published as a chapter in Technological Forecasting and Social Change [45].

## Appendix A: Mathematical Derivations

The following derivations are due to Coleman [13–15]. A collection of n actors exercise control over m goods. This is expressed in a C matrix dimensioned n by m. Control by a given actor is normalized to 1.00 without loss of generality. Similarly, actors have interest in a set of goods which may not be the same goods over which they have control. This matrix is represented by X, and for convenience is transposed and therefore dimensioned m by $n$. We may further scale these matrices without loss of generality, so that the sum total of control across each actor sums to 1.00, and the sum of control of interest across each good also sums to 1.00.

The utility of each actor for receiving control over goods is expressed in the following equation, where utility is an $n$ by 1 matrix (Eq. 9.1). For notational convenience we suspend the subscript on n actors, noting that the same optimization problem applies to each actor. Elements of the matrix are represented using subscripted, lower case letters ($c$ and $x$ respectively).

$$U = \prod_{i=1}^{m} c_i^{x_i} \tag{9.1}$$

We hypothesize a final set of market values, determining the final exchange valuation of each good. This matrix, V is an m by 1 matrix. Actors have resources R which are proportional to the final valuation of their goods times their control. We may now cast the decision problem of the actors as follows:

$$\begin{gathered} \text{maximize (U),} \\ \text{with respect to c} \\ \text{subject to R} = \text{cv} \end{gathered} \tag{9.2}$$

All actors in the system maximize their utility with respect to their decision to exchange control with other actors. However they are subject to a budget, as they are limited to a sum total of exchanges which are equal to their resources. This is reflected in the budget constraint of Eq. (9.2).

The problem may be solved using a Lagrangian, as shown below in Eqs. (9.3a–c). The problem reduces to m equations (one for each actor), plus an additional equation to calculate the Lagrangian multiplier Eq. (9.3c).

$$L = U + \lambda(r - c_i v_i) \tag{9.3a}$$

$$\frac{\partial L}{\partial c_i} = \frac{x_i}{c_i} U - \lambda v_i \tag{9.3b}$$

$$\frac{\partial L}{\partial \lambda} = r - c_i v_i \tag{9.3c}$$

The optimization equation implies that, at equilibrium exchange, the ratio of the marginal utilities is equal to the ratio of going market rates for the good ($v$). Additional linear algebra calculations allows further derivation of the following equations. Equation (9.4a) shows how the stationary value of $r$ is a function of actor control and interest, and Eq. (9.4b) shows the equivalent for market rates. Full derivations of these standing equations are available in Coleman [14].

$$r = \text{CXr} \tag{9.4a}$$

$$v = \text{XCv} \tag{9.4b}$$

Marsden further elaborated the model to include network constraints of trade, where the matrix A is an n by n matrix indicating the social structure of the exchange network. Trades permitted by the network structure of the model are indicated by a 1 in the matrix; trades not permitted by network structure are indicated by a 0. This model too has a potential Markov chain solution Eqs. (9.5a and b).

$$r = \mathrm{rA} \tag{9.5a}$$

$$r = \mathrm{CXrA} \tag{9.5b}$$

Also determined by these equations is the exchange rate (v) for goods in the political or economic exchange. This is the dual problem to determining individual actor resources. As noted earlier, this exchange rate is significant across a number of models of group-decision analysis. Worked example below provides additional mathematical details about the solution of these Eqs. (9.4a, b and 9.5a, b).

As a side note, it is interesting to note that Web search engines calculate the significance of any given page in terms of its "exchange" of hyperlinks with other significant pages on the Internet. This model, embodied in the Google search engine, is fundamentally similar to the Coleman and Marsden models [46].

## Appendix B: Worked Example

Tables A.1 and A.2 show a hypothetical control matrix for 12 microelectronics firms. This corresponds to the C matrix in standard exchange models. Quantities in the table are normalized by column, so that for instance the sum total of "wireless expertise" is summed to 100 %. Quantities in this table might be estimated by research and development indicators (such as patenting).

Table A.3 shows a hypothetical interest matrix for 12 microelectronics firms. This corresponds to the X matrix in standard exchange models. Note that the table, as shown, is transposed. Quantities in the table are normalized by row, so that for instance the sum total of interest of alliance partner "A" is summed to 100 %. Quantities in this table might be estimated through interviews, yearly reports, or industrial classification schemes. A traditional MCDA approach might also be incorporated here.

The example in Sect. 9.4 was run both with complete access to partners, and with limited access to partners. Table A.4 shows a hypothetical alliance structure for this analysis.

**Table A.1** Exchange participants

| Alcatel-Lucent | Nortel networks |
|---|---|
| Ericsson | Royal Philips electronics |
| Matsushita electric industrial | Sagem communication |
| Micron technology | Samsung |
| Motorola | Siemens |
| Nokia corporation | Sony |

**Table A.2** Control matrix (matrix C)

| Firm | Wireless | Micro-electronics | Marketing |
|---|---|---|---|
| A | 0.104 | 0.005 | 0.007 |
| B | 0.000 | 0.071 | 0.229 |
| C | 0.015 | 0.196 | 0.025 |
| D | 0.160 | 0.095 | 0.008 |
| E | 0.008 | 0.092 | 0.177 |
| F | 0.167 | 0.137 | 0.011 |
| G | 0.070 | 0.000 | 0.057 |
| H | 0.186 | 0.000 | 0.174 |
| I | 0.136 | 0.248 | 0.004 |
| J | 0.119 | 0.155 | 0.064 |
| K | 0.023 | 0.001 | 0.095 |
| L | 0.012 | 0.000 | 0.148 |
| Total | 100 % | 100 % | 100 % |

**Table A.3** Interest matrix (matrix X, transposed)

| Firm | Wireless | Micro-electronics | Marketing | Total |
|---|---|---|---|---|
| A | 0.720 | 0.264 | 0.016 | 1.000 |
| B | 0.227 | 0.036 | 0.736 | 1.000 |
| C | 0.558 | 0.294 | 0.148 | 1.000 |
| D | 0.526 | 0.421 | 0.053 | 1.000 |
| E | 0.000 | 0.273 | 0.727 | 1.000 |
| F | 0.578 | 0.266 | 0.156 | 1.000 |
| G | 0.492 | 0.246 | 0.262 | 1.000 |
| H | 0.000 | 0.885 | 0.115 | 1.000 |
| I | 0.292 | 0.053 | 0.655 | 1.000 |
| J | 0.018 | 0.950 | 0.032 | 1.000 |
| K | 0.188 | 0.635 | 0.177 | 1.000 |
| L | 0.665 | 0.332 | 0.003 | 1.000 |

**Table A.4** Alliance structure

| | A | B | C | D | E | F | G | H | I | J | K | L |
|---|---|---|---|---|---|---|---|---|---|---|---|---|
| A | 1 | 0 | 0 | 1 | 0 | 1 | 0 | 0 | 0 | 0 | 1 | 0 |
| B | 0 | 1 | 0 | 0 | 0 | 0 | 0 | 0 | 1 | 0 | 0 | 0 |
| C | 0 | 0 | 1 | 0 | 0 | 0 | 0 | 0 | 1 | 0 | 0 | 0 |
| D | 1 | 0 | 0 | 1 | 0 | 0 | 0 | 0 | 0 | 1 | 0 | 1 |
| E | 0 | 0 | 0 | 0 | 1 | 0 | 0 | 0 | 0 | 1 | 0 | 1 |
| F | 1 | 0 | 0 | 0 | 0 | 1 | 1 | 0 | 0 | 0 | 0 | 0 |
| G | 0 | 0 | 0 | 0 | 0 | 1 | 1 | 0 | 0 | 0 | 0 | 0 |
| H | 0 | 0 | 0 | 0 | 0 | 0 | 0 | 1 | 1 | 0 | 0 | 0 |
| I | 0 | 1 | 1 | 0 | 0 | 0 | 0 | 1 | 1 | 1 | 0 | 0 |
| J | 0 | 0 | 0 | 1 | 1 | 0 | 0 | 0 | 1 | 1 | 0 | 1 |
| K | 1 | 0 | 0 | 0 | 0 | 0 | 0 | 0 | 0 | 0 | 1 | 0 |
| L | 0 | 0 | 0 | 1 | 1 | 0 | 0 | 0 | 0 | 1 | 0 | 1 |

**Table A.5** Alliance structure (matrix A)

| | A | B | C | D | E | F | G | H | I | J | K | L |
|---|---|---|---|---|---|---|---|---|---|---|---|---|
| A | 0.25 | 0.00 | 0.00 | 0.25 | 0.00 | 0.25 | 0.00 | 0.00 | 0.00 | 0.00 | 0.25 | 0.00 |
| B | 0.00 | 0.50 | 0.00 | 0.00 | 0.00 | 0.00 | 0.00 | 0.00 | 0.50 | 0.00 | 0.00 | 0.00 |
| C | 0.00 | 0.00 | 0.50 | 0.00 | 0.00 | 0.00 | 0.00 | 0.00 | 0.50 | 0.00 | 0.00 | 0.00 |
| D | 0.25 | 0.00 | 0.00 | 0.25 | 0.00 | 0.00 | 0.00 | 0.00 | 0.00 | 0.25 | 0.00 | 0.25 |
| E | 0.00 | 0.00 | 0.00 | 0.00 | 0.33 | 0.00 | 0.00 | 0.00 | 0.00 | 0.33 | 0.00 | 0.33 |
| F | 0.33 | 0.00 | 0.00 | 0.00 | 0.00 | 0.33 | 0.33 | 0.00 | 0.00 | 0.00 | 0.00 | 0.00 |
| G | 0.00 | 0.00 | 0.00 | 0.00 | 0.00 | 0.50 | 0.50 | 0.00 | 0.00 | 0.00 | 0.00 | 0.00 |
| H | 0.00 | 0.00 | 0.00 | 0.00 | 0.00 | 0.00 | 0.00 | 0.50 | 0.50 | 0.00 | 0.00 | 0.00 |
| I | 0.00 | 0.20 | 0.20 | 0.00 | 0.00 | 0.00 | 0.00 | 0.20 | 0.20 | 0.20 | 0.00 | 0.00 |
| J | 0.00 | 0.00 | 0.00 | 0.20 | 0.20 | 0.00 | 0.00 | 0.00 | 0.20 | 0.20 | 0.00 | 0.20 |
| K | 0.50 | 0.00 | 0.00 | 0.00 | 0.00 | 0.00 | 0.00 | 0.00 | 0.00 | 0.00 | 0.50 | 0.00 |
| L | 0.00 | 0.00 | 0.00 | 0.25 | 0.25 | 0.00 | 0.00 | 0.00 | 0.00 | 0.25 | 0.00 | 0.25 |

The matrix is normalized so that each alliance partner spends a proportional amount of trading with each of its peers. This is the matrix used for the exchange analysis (Table A.5).

Calculations of stable exchange rates and actor resources proceeds as discussed previously in mathematical derivations above. The resultant eigenvalue problem may be solved using the power method. Since all three of the matrices (C, X, A) are interpretable as probabilities, the problem may be formulated as a Markov chain, and then solved for an equilibrium vectors using a linear system of equations. This is the approach used herein.

Rearranging the equations derived from the actor decision problem (Eq. 9.6a), we have the resultant linear system of Eq. (9.6b), subject to the constraint that the stable probability vector must sum to 1 (Eq. 9.6c).

$$r = \mathrm{CXr} \tag{9.6a}$$

$$(I - \mathrm{CX})\, r = 0 \tag{9.6b}$$

$$\sum r = 1 \tag{9.6c}$$

## References


1. Von Neumann J, Morgenstern O (1944) The theory of games and economic behavior. Wiley, New York
2. French S (1989) Decision theory. Ellis Horwood, Chichester
3. Keeney RL, Raiffa H (1993) Decisions with multiple objectives: preferences and value tradeoffs. Cambridge University Press, Cambridge
4. Luce RD, Raiffa H (1957) Games and decisions: introduction and critical survey. Dover Publications, New York
5. Nash JF (1950) The bargaining problem. Econometrica 18:155–162

6. Nash JF (1949) Two-person cooperative games. Econometrica 21:128–140
7. Kahan JP, Rapoport A (1984) Theories of coalition formation. Lawrence Erlbaum, Hillsdale
8. Osborne MJ, Rubinstein A (1994) A course in game theory. MIT Press, Cambridge
9. Samuelson PA (1938) A note on the pure theory of consumer's behavior. Econometrica 5(7):61–71
10. Varian H (2005) Revealed preference. In: Szenberg M (ed) Samuelsonian economics and the 21st century. Oxford University Press, Oxford
11. Belton V, Stewart TJ (2002) Multiple criteria decision analysis: an integrated approach. Kluwer Academic Press, Boston
12. Keeney RL, Raifa H (1993) Decisions with multiple objectives: preferences and value tradeoffs. Cambridge University Press, Cambridge
13. Coleman JS (1972) Systems of social exchange. J Math Sociol 2:145–163
14. Coleman JS (1973) The mathematics of collective action. Aldine, Chicago
15. Coleman JS (1990) Foundations of social theory. Harvard University Press, Cambridge
16. Marsden PV (1981) Models and methods for characterizing the structural parameters of groups. Soc Netw 3:1–27
17. Marsden PV (1983) Restricted access in networks and models of power. Am J Sociol 88: 686–717
18. Stokman FN, Van den Bos JMM (1992) A two-stage model of policy making with an empirical test in the U.S. energy policy domain. In: Moore G, Whitt JA (eds) The political consequences of social networks. JAI Press, Greenwich, pp 219–253
19. Stokman FN, Zeggelink EPH (1996) Is politics power or policy oriented? A comparative analysis of dynamic access models in policy networks. J Math Sociol 2(1–2):77–111
20. Knoke D (1992) Networks of elite structure and decision making. Sociol Methods Res 22:23–45
21. Sharpf F (1997) Games real actors play: actor-centered institutionalism in policy research. Westview Press, Boulder
22. Koppenjan J, Klijn EH (2004) Managing uncertainties in networks: a network approach to problem solving and decision making. Routledge, London
23. Zaheer A, Bell GG (2005) Benefiting from network position: firm capabilities, structure holes, and performance. Strateg Manage J 26:806–825
24. Oh H, Labianca G, Chung M-H (2006) A multilevel model of group social capital. Acad Manage Rev 31(3):569–582
25. Stuart TE (2000) Interorganizational alliances and the performance of firms: a study of growth and innovation rates in a high-technology industry. Strateg Manage J 28(8):791–811
26. Baum JAC, Calabrese T, Silverman TBS (2000) Don't go it alone: alliance network composition and startups' performance in Canadian biotechnology. Strateg Manage J 21 (3):267–294, special issue
27. Lee GK (2007) The significance of network resources in the race to enter emerging product markets: the convergence of telephony communications and computer networking, 1989–2001. Strateg Manage J 28:17–37
28. Maurer I, Ebers M (2006) Dynamics of social capital and their performance implications: lessons from biotechnology start-ups. Adm Sci Q 51:262–292
29. Schneeweiss C, Zimmer K (2004) Hierarchical coordination mechanisms within the supply chain. Eur J Oper Res 153(3):687–703
30. Schneeweiss C, Zimmer K, Zimmermann M (2004) The design of contracts to coordination operational interdependencies within the supply chain. Int J Prod Econ 92(1):43–59
31. Arrow KJ (1951) Social choice and individual values. Wiley, New York
32. Bots PWG, Hulshof JAM (2000) Designing multi-criteria decision analysis processes for priority setting in health policy. J Multi-Criteria Decis Anal 9(1–3):56–75
33. Shoven JB, Whalley J (1992) Applying general equilibrium. Cambridge University Press, Cambridge
34. Timmermans J (2004) Purposive interaction in multi-actor decision making: operationalizing Coleman's linear system of action for policy decision support. Eburon, Delft

35. Laffont J-J, Martimont D (2002) The theory of incentives: the principal–agent model. Princeton University Press, Princeton
36. Cottle R, Pang J-S, Stone RE (1992) The linear complementarity problem. Academic Press, Boston
37. Brandenburger AM, Nalebuff BJ (1995) The right game: use game theory to shape strategy. Harvard Business Review, July–August, 57–71
38. Mueller DC (2003) Public choice III. Cambridge University Press, Cambridge
39. Lempert RJ, Popper SW, Bankes SC (2004) Shaping the next one hundred years: new methods for quantitative, long-term policy analysis. RAND Corporation, Santa Monica
40. Thoyer S, Morardet S, Rio P, Simon L, Goodhue R, Rausser G (2001) A bargaining model to simulate negotiations between water users. J Artif Societies Soc Simul 4(2). http://jasss.soc.surrey.ac.uk/4/2/6.html
41. Engelman M (2008) Dynamic value networks: position, role and performance of TNO in the semiconductor industry. Delft University of Technology, Delft
42. Lynn B (2006) End of the line: the rise and coming fall of the global corporation. Currency, New York
43. Kakade SM, Kearns M, Ortiz LE (2004) Graphical economics. In: Shawe-Taylor J, Singer Y (eds.), Proceedings of the conference on learning theory, COLT, Banff, Canada, July 2004
44. Bouhtou M, Diallo M, Wynter L (2003) Capacitated network revenue management through shadow pricing. In: Proceedings lecture notes in computer science, vol 2816, pp 341–353
45. Cunningham SW, van der Lei TE (2000) Decision-making for new technology: a multi-actor, multi-objective model. Technol Forecast Soc Chang 76:26–38
46. Langville AN, Meyer CD (2004) Deeper inside page rank. Internet Math 1(3):335–380

# Chapter 10
# Evaluating and Communicating the Value of Technology: Concept and Practice

**Clare Farrukh, Rob Phaal, Letizia Mortara and David Probert**

**Abstract** Evaluating and communicating the value of technology are key considerations for successful innovation. The critical issue of how to evaluate technology as it progresses from exploratory to more mature forms is explored in this chapter, proposing an iterative approach that encourages the alignment of technology and commercial perspectives. Management tools and techniques that can support technology evaluation are discussed, shifting from more qualitative exploratory methods at the start, to more quantitative and analytical tools as developments mature towards commercialisation. Related to this is the other critical issue of how to more effectively communicate the potential value of technology to a wider group of stakeholders, including commercial functions, senior management, funding agencies and external partners. A 'technology marketing' process is proposed, to seek out and identify suitable application opportunities for a promising technology, and to engage with potential partners using a consultative sales framework. The technology evaluation and marketing processes are positioned against the wider backdrop of technology, application and market development and evolution, as an integrating framework. The practical relevance of the approach is illustrated with company case examples.

Adapted from two papers presented at PICMET 2011 (one has since been published in the new IEEE annual book series on Recent Advances in Technology and Innovation Management edited by Joseph Sarkis for Wiley, 'Towards a process framework for assessing the potential value of technologies', the other has been accepted for a special edition of TFSC edited by Tugral Daim, 'The process of making the business case: a sales and marketing perspective for technologists')

C. Farrukh (✉) · R. Phaal · L. Mortara · D. Probert
Department of Engineering, Centre for Technology Management,
Institute for Manufacturing, Cambridge University, Alan Reece Building, 17,
Charles Babbage Road, Cambridge CB3 0FS, UK
e-mail: cjp22@cam.ac.uk

D. Cetindamar et al. (eds.), *Strategic Planning Decisions in the High Tech Industry*, DOI: 10.1007/978-1-4471-4887-6_10,

## 10.1 Introduction

The recognition that technological development is an important driver of innovation, change and growth is not new [1]; however, it is now widely accepted that this is increasingly true for enterprises of all sizes in all sectors [2, 3]. Promotion of technology-based innovation is a major policy initiative in most advanced economies.

Whether or not to invest in technology is thus a fundamental question for most technology-based firms. Two key questions managers and technologists must address are '*what* is the value of this technology?' and consequently '*how* to communicate the value of this potential asset?' This requires consideration of what appraisal techniques should be used, what criteria are relevant for decision making? And what process related issues should be taken into consideration? These are challenging questions, with the answers depending on contextual factors such as the maturity level of the technology, the degree of uncertainty that surrounds the particular valuation question, and the identity of the individuals (i.e. internal stakeholders or external investors and analysts) wishing to establish this value.

There is a need for an integrated process approach to evaluating technology potential and to building a business case in a way that facilitates communication between technologists as 'sellers' and internal or external customers as 'buyers'. Different perspectives that might be considered include those of a national government funding central R&D labs, a consortium of firms engaged in joint development projects, individual firms acquiring or exploiting new technology, or even financiers wishing to invest in a company based on the development of a new technology.

The contributions of this chapter can be positioned against the diagram shown in Fig. 10.1. Successful innovation requires an alignment between technology (resources), application (products and services) and market (customer needs and competitive environment), balancing technology push (supply) with market pull (demand). These three perspectives are combined in Fig. 10.1, illustrating the iterative logic used to align technological and commercial perspectives in innovation and strategy, based on road mapping principles [4, 5]. Confidence in technology investments requires effective approaches for exploring and assessing the future potential value that the technology might enable in the market.

This chapter first sets out a framework for iterative technological evaluation as technology progresses from exploratory to more mature forms along the development funnel, including consideration of appropriate management tools and techniques (Sect. 10.2). Such technological evaluation requires consideration of technological, application and market considerations, appraising current investments with the potential for future revenues. A process is then proposed for developing the business case for technology, based on marketing and sales concepts to help to identify and communicate the value of technology to internal commercial stakeholders and external customers and investors (Sect. 10.3). Technology evaluation is a key component of this process.

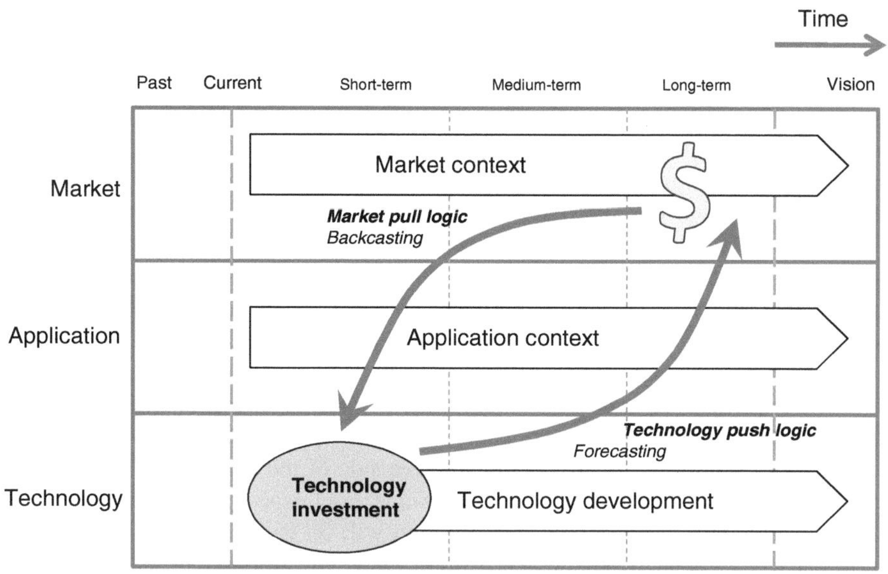


Fig. 10.1 Aligning technology, product and market perspectives for innovation and strategy

In the following sections, key concepts and issues that surround the assessment and communication of technology investments are discussed. The unit of analysis for valuation is the underlying technology, which may be at varying stages of maturity, from very early stage research through to implementation in the market place, typically embodied in a product, service, system or process.

## 10.2 How to Evaluate Technology

There are several aspects to consider in building a framework for the evaluation of technology that go wider than purely financial considerations, especially at early stages of technology and market development where hard data are scarce. These include the maturity of the technology and the nature of the process of evaluation.

### *10.2.1 The Role of Technology Readiness*

The evaluation challenge has distinct characteristics throughout the lifecycle of technology projects which may be considered in three main stages, namely the early stage technology, mid stage technology and late stage technology. This is, however, a generalisation and the timeline may be broken into different stages in different industrial sectors. For example, in pharmaceuticals the stages can be early drug discovery, the pre-clinical phase and the clinical phase, which in turn are separated into three or four sub-phases. In some other sectors, for example aerospace, the term technology readiness levels (TRL) is used [6]. TRLs separate the 'readiness' or maturity of a technology into nine different levels. Regardless of the number of segments used to study the technology lifecycle, three general stages can be identified: proof of concept, proof by demonstration and proof of application, as shown in Fig. 10.2.

As the technology development project progresses the uncertainty about the underlying assumptions of the valuation decreases. However, at the same time, in many cases the investment requirement increases. This has an impact on the selection of the appropriate techniques for valuation. For example, in the proof of concept phase (TRLs 1–3, Early Stage) the investment is often highly uncertain and questions such as "will it actually work?" and "if it works will there be an application and a market?" are commonplace. Appropriate techniques in this phase are more geared towards the qualitative side of the analysis. Whereas in the proof of demonstration (TRLs 4–6, Mid Stage) and especially in proof of application (TRLs 7–9, Late Stage), the valuation will be based on a business case to show clear return on investment potential. Here, the importance of quantitative valuation methods increases.

## *10.2.2 The Nature of the Process of Technology Evaluation*

The use of TRLs suggests that evaluation is an iterative process, as the technology develops and its application environment emerges and develops the cycle of evaluation must be repeated. In addition, there are a series of very different

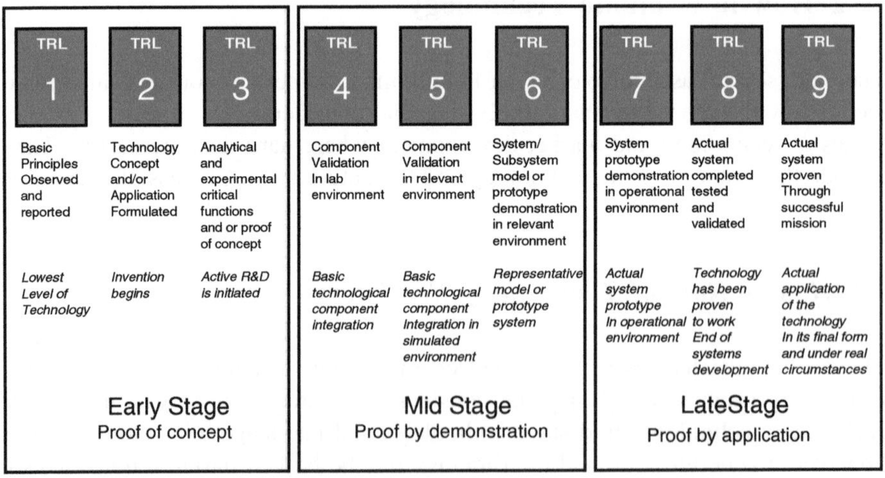


**Fig. 10.2** Generic phases of technology readiness (adapted from [6])

evaluative activities that must be addressed, which may be summarised in three main categories: exploration, quantification and communication:

- *Exploration*: exploring potential value opportunities for new technologies is challenging. Research shows that decision makers often rely on expert judgement and gut feel for making decisions under conditions of high uncertainty and complexity [7, 8].
- *Quantification*: in order to make an investment in technologies, some type of formal quantification of its potential return is required. Discounted cash flow (DCF) techniques are easy-to-use, intuitive, widely applicable, credible and accepted [9]. However, research shows that myopic use of the technique can lead to poor decision making [10], and such techniques are quite unreliable for early stage technology and innovation where forecasts are unreliable, with weighted scoring method typically deployed [11, 12]. Decision tree approaches are one way of prompting 'options thinking' to allow management flexibility to be taken into account [3, 13].
- *Communication*: early involvement of stakeholders results in a common understanding of assumptions and approach and can be achieved by means of workshops. Visualisation techniques include road mapping [14], portfolio management [12, 15] and consultative sales [16, 17].

The inputs to the technology valuation process are often a combination of ideas, interpretations, expert judgements and gut feel and intuition. These include ideas regarding new technologies, interpretations of how the technology can be developed and how it can be applied and the value associated with this application, and expert judgements about whether it has a chance of success based on previous experience. Furthermore, these inputs are either aligned with the strategy or from the basis of a new strategy, and are complemented with available factual data when the act of valuation commences. These observations have been developed into a generic process integrating the most relevant elements (Fig. 10.3).

### *10.2.3 Evidence from Practice*

Sixteen semi-structured interviews were carried out to explore valuation practices in the UK-based companies (Table 10.1), covering a range of sectors and company sizes, ranging from large firms (>10,000 employees) to small enterprises (<250 employees).

Table 10.2 provides an overview of the common practices, techniques and tools as identified in selected sectors for the companies involved in the research. These have been classified into the three areas identified in the literature review: exploration, quantification and communication, at the various stages of technology maturity. The interview results did not reveal any particular differences between large and small companies in each sector, but this may have been due to the small numbers of companies in the sample, and in particular the lower proportion of

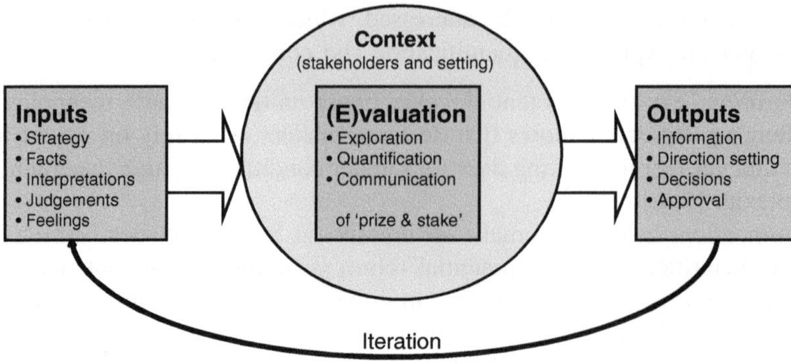


**Fig. 10.3** Technology valuation as an iterative process of exploration, quantification and communication

**Table 10.1** Company interviews—technology evaluation

| Sector | Number of companies | Company size |
|---|---|---|
| Aerospace | 4 | Large |
| Telecoms | 1 | Large |
| Consumer electronics | 3 | 2 large, 1 small |
| Scientific instruments | 3 | Small |
| Printing | 2 | 1 large, 1 small |
| Food | 1 | Large |
| Automotive | 1 | Large |
| Pharmaceutical manufacturing | 1 | Large |

small firms involved. Comparing sectors, the aerospace, consumer electronics and scientific instrument companies did appear to use a wider range of tools than the automotive, printing and pharmaceutical companies, which might suggest less awareness or need for valuation tools, but may also have been due to the product and manufacturing focus of the interviewees. However, the results did not show that companies' choice of tools is affected by technology maturity, and this is explored further below.

#### 10.2.3.1 Proof of Concept Phase (Early)

In the proof of concept phase, the emphasis is predominantly on exploring value opportunities and scoping out options, and communicating these options accordingly. As uncertainty is very high at this stage, and the investments are relatively low compared to the later proof by demonstration and proof by application phases, calculation is a low priority. Calculations in this phase can support the development of decision trees and options analysis, but have to be presented with a proportionate respect to the underlying uncertainty. As seen in the consumer

**Table 10.2** Common practices, tools and techniques (selected sectors)

| Sector | Stage | Early | Mid | Late |
|---|---|---|---|---|
| Aerospace | *Exploration* | • Gut feel/expert judgement<br>• Roadmapping<br>• Networking (e.g. matching funds) | • Roadmapping<br>• Scoring systems | |
| | *Quantification* | • Expert judgement<br>• Monte Carlo<br>• Portfolios and scoring | • Hybrid DCF<br>• Portfolios | • Finance based portfolios<br>• Business cases, ROI for options, payback |
| | *Communication* | • Informal discussions with technical experts<br>• Lobbying | • Formal presentations<br>• Direct negotiations | • Formal presentations<br>• Direct negotiations |
| Automotive manufacturing | *Exploration* | • Expert judgement in technology | • Demonstrators<br>• Marketing intelligence (e.g. market driven features)<br>• Brainstorm workshops | |
| | *Quantification* | | • Planning<br>• Car line driven quantification | • Strategic innovation plan |
| | *Communication* | • Presentations/'shopping trolley'<br>• Lobbying | • Informal discussions/lobbying<br>• Demonstrators<br>• Stage gates<br>• Aligning product and technology groups | • Formal process<br>• Stage gate driven<br>• Lobbying |
| Consumer electronics | *Exploration* | • Technology intelligence and scouting | • Informal<br>• communications<br>• Roadmaps<br>• Portfolio management<br>• Steering team | |

(continued)

**Table 10.2** (continued)

| Sector | Stage | Early | Mid | Late |
|---|---|---|---|---|
| | *Quantification* | • Gut feel market size | • Stage gate review | • Market size, time to market, novelty, competition and margins |
| | | | • Market size, time to market, novelty, competition and margins | |
| | *Communication* | • Lobbying | • Visualisation of idea in innovation funnel | • Mock ups and pitch |
| | | • Informal networking (with HQ) | • Internal reviews | • Informal networking before formal presentation and negotiation |
| | | | • Networking,experience | |
| | | • Presentations at weekly meeting | | |
| Printing | *Exploration* | | • Product driven exploration | |
| | *Quantification* | | • Product business case | • Product business case |
| | | | • Only product driven investments in technologies | • Only product driven investments in technologies |
| | *Communication* | | • Informal lobbying | • Board meetings |
| Pharma manufacturing | *Exploration* | • Scouting | • Roadmaps | |
| | *Quantification* | • Budgets | • DCF moving to decision trees and Monte Carlo simulation | • Business cases, DCF |
| | *Communication* | • Boards, meetings and informal/formal presentations | • Roadmaps, decision trees | • Business case reviews by division heads |

electronics sector, gut feel informed by the results of technology intelligence and scouting activities are used to estimate market size at this early stage.

For example, for early stage technologies in an aerospace manufacturer, the input to the process was based on strategy, facts, interpretations and judgements. In blue sky technology development projects (early stage), the Technology Director revealed that appropriate alignment of strategy at the beginning of the process was fundamental. Strategy in this sense does not only refer to the company but also to the strategy of customers (in this case the original equipment manufacturer, or OEM) and government funding agencies (as these funding agencies provide significant investment into new technology developments). The strategy sets the stage for the valuation process. Exploration is done by engaging senior experienced engineers. This was reported to be an informal process where the interpretation of facts and specific judgements are collected over a period of time. The final decision to invest is made based on discussions with the Technology Director and the Chief Executive Officer. For early stage investments few formal calculation tools were applied. As the Technology Director aptly pointed out, if he got it wrong it was his problem and accountability, and manipulating numbers in a financial way did not change this fact.

#### 10.2.3.2 Proof by Demonstration Phase (Mid)

In the proof by demonstration phase, the application and the selection of what concept will go forward to application becomes critical. Exploration is still important but this phase is focused on finding a balanced approach with respect to the existing portfolio from a strategic point of view. Sharing of multifunctional perspectives is crucial for a balanced evaluation (e.g. marketing, technology and finance). Establishing a discourse and reaching a common understanding is key in this phase in order to reach a decision to approve pursuing or stopping the development. The review of practice revealed that stopping projects at this stage seems particularly difficult when prior commitment has already been approved. Hence, integrating these perspectives with appropriate jointly constructed evaluations is particularly relevant, often supported by visual depictions (e.g. roadmaps).

For example, in a pharmaceutical company, road mapping has been applied as an integrating technique for aligning the various perspectives of different departments for investments in new manufacturing technologies. Other techniques that have been observed in this phase are the use of options based decision trees which encompasses both exploration and visualisation. Our research shows that selection of the right approach and technique is not always done satisfactorily and reinforces the argument to have an integrative approach.

An aerospace manufacturer highlighted the importance of alignment along the timeline. A team of technology managers was requested to put forward a comprehensive business case for a new technology, with the potential applications justified by financial calculations. However, the technology was still in a proof of concept stage, with high levels of uncertainty. The setting in which the business case was presented was a 15 min pitch to senior decision makers in dedicated meetings held at 6-month intervals. The anticipated result of the presentation was a go or no-go decision. However, as the uncertainty was high, and potential intermediate steps had not been explored (e.g. using decision trees and real options), an investment decision could not be made. Further evaluation was requested for review at the next meeting. This was repeated six times over a period of 3 years, with the business case focussed on obtaining full funding for the technology development (proof by demonstration), rather than phasing the programme and funding. As no funding was received over a period of 3 years the company risked losing advantage over the competitors. Resources were allocated to refine the presentation instead of comparing the decision options available. The company subsequently reviewed and adapted their valuation process to avoid this trap. This example shows that disconnects between the purpose of the valuation (e.g. exploration, calculation or communication) and the selected context, setting and stakeholders, can result in an unsatisfactory valuation exercise and indicate the need for process improvements.

In the case of a car manufacturer, the use of early demonstrators was reported to be an appropriate means for gaining the interest of key decision makers. Low cost demonstrators were prepared (e.g. 2D models) and funded from the existing internal budgets, and used in informal meetings with decision makers to show the potential application of the technology. The impression from the innovation group responsible for linking new technologies to new product features was that this had a positive impact on the valuation process. A similar example was observed in the case of a scientific instrument supplier using low cost demonstrators to illustrate potential applications. This example shows that the use of a deliberate, albeit informal, communication setting with decision makers early in the valuation process is important to the overall process.

#### 10.2.3.3 Proof by Application Phase (Late)

In the proof by application phase, the technical uncertainty is typically significantly less and the market projections are more apparent. Calculating future returns using DCF and real options techniques becomes more relevant. However, it is still important to properly communicate and get agreement on these projections, and to involve the right stakeholders.

An example from a printer manufacturer showed a decision process that led to the elimination of a technique used to select the right ideas from a portfolio of projects. The company only invested in technologies that could be directly applied. The board responsible decided to use a portfolio method to help them shape judgement on what to invest in. After the internal exercise, the general opinion was that the technique did nothing more than visualise what they already thought. The chosen technique (portfolio management) and setting (board meeting with decision makers) did not work well together. The portfolio management technique was

appropriately used to visualise the portfolio. However, the problem was that the group involved already knew and had a consensus on this picture even though it was never made explicit. A disconnect between context, purpose, stakeholders and chosen approach resulted in the rejection of the technique in future valuation processes.

In this phase, the emphasis is much more geared to having appropriate financial justification and projections (e.g. on potential return on investment). The aerospace manufacturer that used a very informal approach for early stage technology valuation also used a comprehensive financial model to balance projects in the portfolio when it came to the proof by application stage. Reporting, justifying and communicating investment decisions were also more significant at this stage. Another example from an aerospace OEM was the use of a wide range of techniques (e.g. predictive modelling software) to get as many perspectives as possible on the potential outcome.

### *10.2.4 An Integrative Process Approach*

For an integrative approach to technology evaluation, it is thus important to acknowledge the need for a range of techniques, as different tools and criteria are appropriate at different stages. Furthermore, the case studies show that technology development is not always a linear process, with a need to include iteration, learning and feedback. This is depicted in Fig. 10.4, which combines the iterative loops of Fig. 10.3 (input-evaluation-output) with the innovation funnel. The funnel depicts the reduction of options and alternatives as the technology matures; however, the level of investment typically increasing as uncertainty decreases. This has a profound impact on which approaches are appropriate, and how they should be used in each phase, evolving to become more quantitative as technology moves towards being embedded in a product. Whatever the maturity, effective technology evaluation requires a good understanding of the potential 'route to value'.

### *10.2.5 Tools to Support the Evaluation of Technology Potential*

An iterative approach to evaluating technology potential is proposed, with balance between qualitative and quantitative emphasis changing with TRL. The different types of decision support tools that this suggests are shown in Table 10.3.

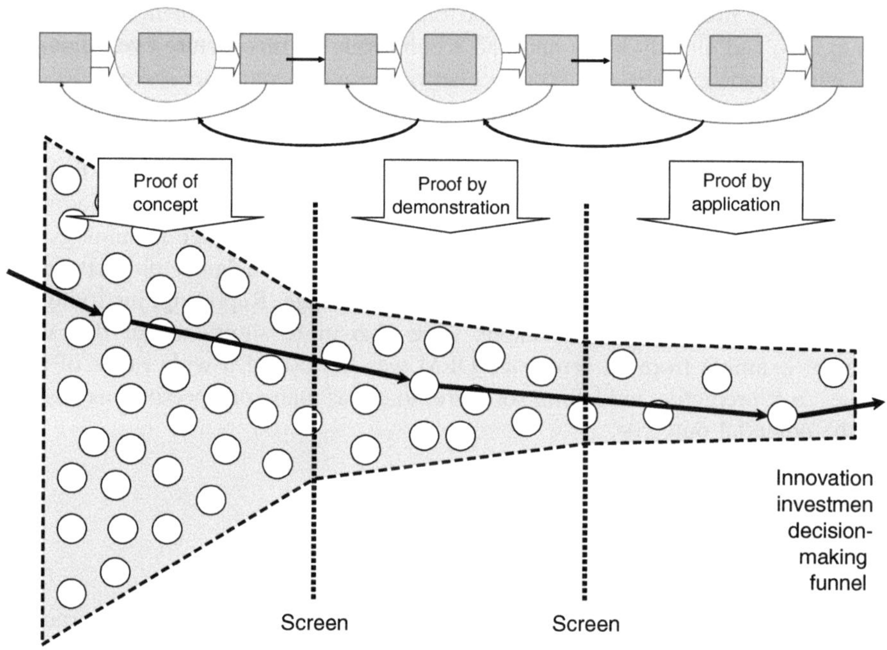


**Fig. 10.4** An integrative phased iterative process approach to technology evaluation

**Table 10.3** Examples of how decision support tools can support technology evaluation for different stages of technology maturity

| | Early stage technology | Mid stage technology | Late stage technology |
|---|---|---|---|
| | TRL 1–3 | TRL 4–6 | TRL 7–9 |
| | Proof of concept | Proof by demonstration | Proof by application |
| | Iteration—explore, quantify and communicate | Iteration—explore, quantify and communicate | Iteration—explore, quantify and communicate |
| *Explore* | Expert judgement and gut feel | Road mapping, scoring and portfolio methods | Linking grids for technology, product and market |
| *Quantify* | Options thinking | Decision trees, scoring and portfolio methods | DCF and net present value |
| *Communicate* | Concept papers, involvement and visualisation | Technology demonstration, workshops, visualisation and demonstrators | Application demonstrators, pilot trials |

## *10.2.6 The Communication Gap*

As discussed above, technological investment is a key driver of innovation and the evaluation of technology potential is becoming increasingly important in this context. A range of approaches and tools are available for developing an

understanding of the value of technology and a process perspective is proposed. However, communicating the potential value of technology to possible customers (internal and external) is not well documented in terms of theory and practice and falls outside the skill set of many technologists. There is potential to learn from more commercial approaches, such as sales and marketing disciplines.

The next section seeks to integrate the concepts of marketing and consultative selling into making the business case for new technology. It describes a five step process model for technologists wishing to build an effective business case for securing investment internally or when 'selling' a technology.

## 10.3 How to Communicate the Potential Value of Technology

New technology developments are often characterised by high levels of both market and technological uncertainty [18, 19]. This complicates the task of obtaining internal funding, for example from business units, or to selling them externally, for example as licenses.

This section focuses on understanding how a marketing and sales perspective can support investment decisions for new technologies either inside the firm or selling technology to external customers. In both cases, it is often the research and development departments that have the task of securing financial support for the new technology, be it from internal sources such as product or service-oriented business units, or from external investors or customers such as those who wish to buy for example, licenses to be able to use the technology in their products or to improve their processes. The sales and marketing expertise within a company is usually focused on selling products or services that forms the main income stream of the business and so it is not generally available to support these technology-focused activities.

### *10.3.1 Defining the Business Case*

A typical description of 'building the business case' in a technology-based company is provided by Cooper [20], who suggests that technical, marketing and business feasibility are assessed and result in a business case with three main components: product and project definition, project justification and a project plan. However, as [21] recognises "…in a technology development project, the commercial prospects for the new technology are often unclear, especially near the beginning of the project when these commitment decisions are required" (p. 24). Thus, building the business case for technology can be seen as an entrepreneurial activity. Reference [22] discusses the literature on entrepreneurial business planning and emphasise the crucial importance of communication in making a business

case. They define an entrepreneurial business plan as "the formal argument used to secure, from prospective investors, resources for a proposed entrepreneurial process".

## *10.3.2 Matching Technology and Opportunity or Problem*

For technologists faced with the situation of promoting a technology that they have been heavily involved in developing, it can be hard to stand back and view it through the eyes of a buyer or customer. However, as Freeman [23] states, "Innovation is essentially a two-sided or coupling activity..." (p. 109). He suggests that a process of 'matching' the technical possibilities and the market may be carried out using experimental development and design, trial production and marketing. This concept suggests that in order to sell a new technology, it is necessary to demonstrate sufficient understanding of potential market opportunities rather than rely on the buyer to make the connection. Hence, technologists might benefit from an insight into the two-way communication process embedded within sales and marketing approaches, in order to develop a 'pitch' or business case for their technology. Several aspects of the literature are relevant here:

- *Marketing and selling of technologies*: there are key differences between selling know-how and a defined product [24, 25]. These include intangibility, unwillingness of technical staff in the buying organisation, distribution, market identification and pricing.
- *Technology adoption*: it is necessary to consider both individual and collective adoption behaviour. The process of adoption and its enablers [26] are important, as are the characteristics of potential buyers of technology over time [27].
- *Technology valuation*: a key input to building the business case, with both qualitative [9, 20] and quantitative [10, 28] approaches available. However, expert judgement or 'gut feel' often plays a significant role [29], especially when reliable data are scarce.
- *Resource allocation process*: the reality of organisational decision making [30] indicates the pre-work necessary for making a successful business case. Dean [31] refers to the need to 'sell' the technology within an organisation and to attend to strategic, financial, interpersonal and political aspects.
- *Consultative sales*: gaining a commercial perspective can be difficult for technologists. Two-way sales processes and techniques, such as clearly understanding a customer's needs and objectives [16], and ensuring that the customer agrees that these needs should be addressed [32], are practical and buyer focused.

The context for this work is a capabilities based framework for technology management activities [33]. The building of the business case itself and the evaluation activities that precede it, support both the acquisition and exploitation of technology as framed by the strategy, innovation and operations processes of the firm. A further objective is to link the commercial and technological

perspectives held within the firm that drive resource allocation in creating and responding to new organisational and environmental opportunities.

### 10.3.3 A Practical Process for Building the Business Case

A process activity model has been developed to support the successful construction and delivery of an effective business case [34]—see Fig. 10.5. The process is aimed at both internal and external technology sales situations and incorporates market research and consultative sales perspectives. Technology evaluation activities (Sect. 10.2) are conducted as part of this process.

The five-step generic process for securing technology investment from a seller's perspective is summarised below. These steps represent the generic market research and sales activities that support a business case, and apply to both internal and external customers for technology:

1. Identify the technology/problem combination

   The first step has been designed to combine market research and technology analysis, and focuses on identifying the key markets or industries that could benefit from the technology. This involves actively seeking opportunities and problems for which the technology can produce a benefit. The technology/problem combinations may be established using methods such as market segmentation and by identifying potential customers, both internally and externally.

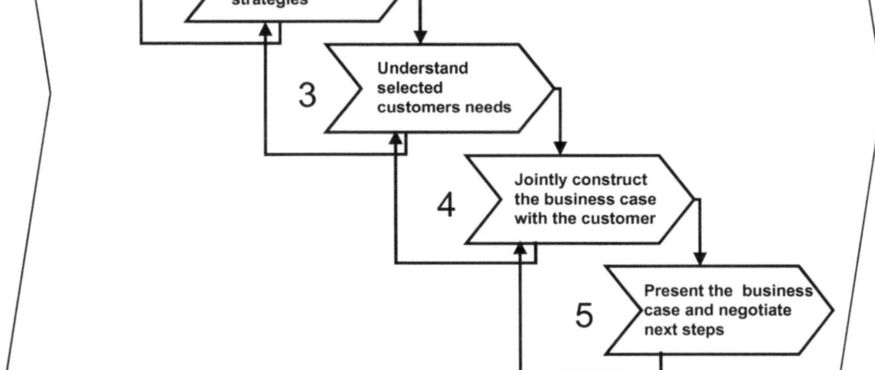


**Fig. 10.5** Five steps for building a business case for technology [34]

2. Select potential customers and sales strategies

   This step aims to find out which customers have a clear need for the technology. Typical criteria for customer selection include the benefit and value of the technology/problem combination, availability of the necessary funds, and whether the individual has the authority to make the investment.

3. Understand selected customers' needs

   Here, the emphasis moves from market research towards building a deep understanding of the needs of the customers. This involves finding out who is actually in the 'buying centre', as this is not necessarily always the user of the technology (e.g. engineering or R&D department). This is the stage where the relationship is built between seller and customer. It is important that the seller listens carefully to find out whether the customer's problems might be solved by their technology. It is more important for the seller to understand the customer at this point than vice versa.

4. Develop/construct the business case with the customer

   The fourth step involves a joint effort by both seller and customer to develop the business case to meet the customer's needs. The dialogue that emerges should eliminate any mismatch between the technology and the problem: if one party feels they have been misunderstood, the process can be fed back to the previous step. A jointly constructed business case prevents divergent expectations during its final presentation.

5. Present the business case and negotiate next steps

   The last step is to present the business case for a go/no-go decision. By this time, the seller is well prepared and buy-in has already been established. The seller knows exactly what the customer requires and provides an unambiguous and understandable business case based on these requirements.

Technology valuation activities are seen as both a trigger and an input to the process. The early realisation of the potential value of a technology may be less based on tools and techniques and more on insight and gut-feel, but a subsequent 'matching' of technology and opportunity is the start of the business case journey [23]. Later on, the use of more formal techniques for quantifying and visualising value form part of the business case preparation, giving the crucial 'depth' to the argument [31]. The maturity of the technology affects the degree of uncertainty that has to be dealt with. The nearer the technology is to application or market, the easier it is to construct the business case. The concept and practice of 'demonstration' is useful throughout the process, to support communication and alignment, and as a vehicle to reduce uncertainty [5].

At the start, it is worth reviewing what knowledge, competence, software, copyright or patents are involved. As the process moves to considering potential partners, it is necessary to be aware of the danger of exposing IP to potential customers before agreement is reached in principle. Ultimately, however, the way the IP is packaged needs to be acceptable to the buyer. If the final presentation is made externally, confidentiality issues need to be addressed.

The political context of the process is implicit in its structure, primarily in the joint preparation of the business case with the customer/buyer. This seeks to ensure that many of the necessary clarifications and negotiations take place as part of the process rather than being raised at the final 'pitch'. For example, as part of understanding the needs of the selected customers, it is important to involve all internal stakeholders and the most significant external interested parties in the discussion process.

### *10.3.4 Evidence from Practice*

Six semi-structured interviews were carried out to explore technology marketing and selling practices in the UK-based companies (Table 10.4), with key findings summarised in Table 10.5.

The five steps are illustrated for one of the cases in Fig. 10.6, for a multinational organisation with headquarters in the USA and subsidiaries worldwide. The company had invested substantially in building and improving their technology intelligence system, which designates certain staff to *identify* (or 'scout') new opportunities. It is the scout's responsibility to develop initial business cases for interesting opportunities, matching technological capability with market need. These are presented and opportunities reviewed at weekly meetings. Individual champions are designated for interesting opportunities, responsible for further investigation to gather evidence about the technology and its potential.

In parallel, internal networking is undertaken to establish if and how the technology might be adopted and exploited, and to identify internal stakeholders and *select* potential business champions ('customer'). Further discussion and investigation deepens *understanding* of the opportunity and the needs of the business, including enablers and barriers to adopting the technology within the organisation. Demonstration forms a key part of *constructing* the business case for what can be a substantial investment, developed in collaboration with the business champion. This is tested informally before the business case and associated demonstration is *presented* to senior decision makers in a formal 30 min 'pitch'.

**Table 10.4** Company cases: making the business

| Sector | Number of companies | Company size |
|---|---|---|
| Aerospace | 2 | Large |
| Automotive | 1 | Large |
| Imaging | 1 | Large |
| Industry research organisation | 1 | Large |
| Pharmaceutical manufacturing | 1 | Large |

**Table 10.5** Summary of case findings: making the business case

| Company | Technology–what does the company wish to sell either externally or internally? | Process steps and key issues revealed in case study |
|---|---|---|
| Aerospace 1 | External customer: software as technology, 'black box' products<br>Internal customer: decisions to give investment to R&D, universities or SMEs to develop technology to proof of concept stage | The company showed a focus on active selling of technology in the company. Due to well established internal and external customers, market research was not seen as necessary. [Sales emphasis] |
| Industry research organisation (IRO) | External customer: selling technology (know-how) developed alone or in collaboration with partners (e.g. Universities)<br>Internal customer: n/a | The organisation focused on the importance of market research as a precursor to 'selling' activities in developing the business case. [Market research emphasis] |
| Imaging | External customer: n/a<br>Internal customer: when an external technology is interesting for possible investment by BU or R&D; or an internal technology has a stage gate review | The company described a process including a well-developed technology intelligence system. Hence, responsibility for proposing investment in external technology is well defined. [Market research and sales] |
| Aerospace 2 | External customer: non -core IP from R&D<br>Internal customer: n/a | The company discussed its existing process in a workshop and identified areas for improvement. [Market research and sales] |
| Automotive | External customer: n/a<br>Internal customer: new technology inputs to vehicle programmes or production environment | The company sees commercialising technology as a difficult internal process. It involves many influential stakeholders and interaction with established technology, product and financial systems. [Market research and sales] |
| Pharma (manufacturing) | External customer: out licensing of technology no longer required by company or as part of a collaboration, software licences<br>Internal customer: drug delivery technology or manufacturing technology needed after molecular development of drug | The company described a process but with interesting internal and external variation. Externally, a ventures group had recently disbanded. Internally, both the identification of the technology/ problem area and the development of the business case were highly influenced by therapy divisional chiefs. [Market research and sales] |

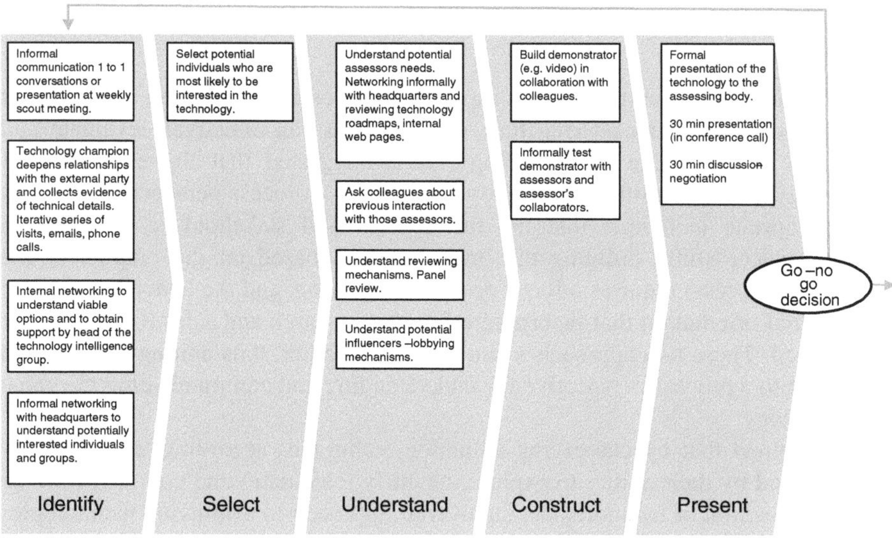


**Fig. 10.6** Example company process for building the business case for investment

Although the technology is of medium maturity, it can be seen that the first four steps of the process are largely informal.

### *10.3.5 Tools to Support the Development of the Business Case*

A range of decision support approaches and tools can support the proposed five-step process for building the business case for new technology investment, as illustrated in Table 10.6, drawing upon the literature and practice.

**Table 10.6** Examples of methods that can support development of the business case

| | |
|---|---|
| (1) Identify the technology/problem combination | Searching, scanning and technology intelligence |
| (2) Select potential customers and sales strategies | Force field analysis, mind maps, databases, score cards, portfolio management and SWOT |
| (3) Understand selected customers' needs | Listening/questioning (open), use of demonstrators and practice in being challenged |
| (4) Develop/construct the business case with the customer | Structuring and visualisation, business model templates, options thinking, demonstrators, project planning, financial projections e.g. DCF/NPV, business model options, 'straw men' presentations |
| (5) Present the business case and negotiate next steps | Presentation skills, demonstrators, closed questions, walk-away price and negotiation skills |

## 10.4 Summary and Conclusions

An integrated approach to two key aspects of successful innovation management has been presented, considering how organisations can effectively evaluate and communicate the value of technology. It is suggested that the evaluation of technology can be significantly improved by using a process perspective to select the appropriate technique that fits the context and stakeholders involved. In addition, successfully building the business case (based on the outputs of the evaluation process) requires a focus on communication, and the benefits of a more commercial orientation that incorporates market research and sales techniques are highlighted. These two approaches can be used together, thus uniting concept and practice into a process perspective for understanding and communicating the value of technology.

It is argued that by classifying valuation techniques according to technology maturity and by their ability to explore, quantify (calculate) and communicate the value proposition of technologies, the overall approach to evaluating technologies for investment decisions can be improved. Combining effective methods for evaluating the potential of technology with a clear five-step process for building a business case to secure technology investment leads to an integrative process for technology evaluation and communication.

For practitioners, the process perspective can be used as a guideline to organise and structure the activities of technology valuation consistently over the life cycle of the technology, using exploration, calculation and communication as building blocks over the three generic phases of technology development. It is not assumed that a generic decision-making process is useful to all, recognising that the design of the process for valuing technologies remains one that needs to be carried out within the specific context and strategy of a firm; the integrated logic provides a guide for R&D managers to utilise the available building blocks in a consistent manner.

## References

1. Goffin K, Mitchell R (2005) Innovation management. Palgrave Macmillian, Basingstoke
2. Smit HTJ, Trigeorgis L (2006) Strategic planning: valuing and managing portfolios of real options. R&D Manage 36(4):401–419
3. Schneider M, Tejeda M, Dondi G, Herzog F, Keel S, Geering H (2008) Making real options work for practitioners: a generic model for valuing R&D projects. R&D Manage 38(1):85–106
4. Phaal R, Simonse L, van Ouden E (2008) Next generation road mapping for innovation planning. Int J Technol Intell Plan 4(2):135–152
5. Phaal R, O'Sullivan E, Routley M, Ford S, Probert D (2011) A framework for mapping industrial emergence. Technol Forecast Soc Chang 78:217–230
6. Mankins JC (2009) Technology readiness assessments: a retrospective. Acta Astronaut 65:1216–1223

7. Pavia TM (1991) The early stages of new product development in entrepreneurial high-tech firms. J Prod Innov Manage 8(1):19–31
8. Bannister F, Remenyi D (2000) Acts of faith: instinct, value and IT investment decisions. J Inf Technol 15(3):231–241
9. Hunt F, Mitchell R, Phaal R, Probert DR (2004) Early valuation of technology: real options, hybrid models and beyond. J Soc Inst Control Eng Japan 43(10):730–735
10. Boer FP (1998) Traps, Pitfalls and Snares in the valuation of technology. Res Technol Manage 41(5):45–54
11. Martino JP (1995) R&D project selection. Wiley, New York
12. Cooper R, Edgett S, Kleinschmidt E (2001) Portfolio management for new product development: results of an industry practices study. R&D Manage 31(4):361–380
13. Faulkner TW (1996) Applying 'options thinking' to R&D valuation. Res Technol Manage 39(3):50–56
14. Willyard CH, McClees CW (1987) Motorola's technology road mapping process. Res Manage, 13–19 Sept–Oct
15. Kahraman C, Ruan D, Tolga E (2002) Capital budgeting techniques using discounted fuzzy versus probabilistic cash flows. Inf Sci 142:57–76
16. Hanan M (1986) Consultative selling: get to know your customers' problems. Manage Rev 75(4):25–31
17. Delvecchio S, Zemanek J, McIntyre R, Claxton R (2004) Updating the adaptive selling behaviors: tactics to keep and tactics to discard. J Mark Manage 20:859–875
18. Naveen Y, Swami S, Pal P (2006) High technology marketing: conceptualization and case study. Vikalpa J Decis Makers 31(2):57–74
19. Maine E, Garnsey E (2006) Commercializing generic technology: the case of advanced materials ventures. Res Policy 35(3):375–393
20. Cooper RG (1986) Winning at new products. Addison-Wesley, Reading MA
21. Cooper RG (2006) Managing technology development projects. Res Technol Manage Nov–Dec, pp 23–31
22. Hindle K, Mainprize B (2006) A systematic approach to writing and rating entrepreneurial business plans. J Private Equity 9(3):7–22
23. Freeman C (1982) The economics of industrial innovation. Frances Pinter Publishers, London
24. Ford D, Ryan C (1977) The marketing of technology. Eur J Mark 11(6):369–382
25. Tschirky H, Escher JP (2000) Technology marketing: a new core competence of technology-intensive enterprises. Int J Technol Manage 20(3–4):459–474
26. Rogers EM (2003) The diffusion of innovation. The Free Press, New York
27. Moore GA (1998) Crossing the chasm. Capstone, Oxford
28. McGrath RG (1997) A real options logic for initiating technology positioning investments. Acad Manag Rev 22(9):974–996
29. Sadler-Smith E, Shefy E (2004) The intuitive executive: understanding and applying 'gut feel' in decision-making. Acad Manage Executive 18(4):76–91
30. Bower JL (1986) Managing the resource allocation process (Revised Edition). Harvard Business School Press, Boston
31. Dean JW (1987) Deciding to innovate—how firms justify advanced technology. Ballinger, Cambridge Mass
32. Graziano JE, Flanagan PJ (2005) Explore the art of consultative selling. J Account, pp 34-37 (January)
33. Cetindamar D, Phaal R, Probert D (2009) Understanding technology management as a dynamic capability: a framework for technology management activities. Technovation 29(4):237–246
34. Dissel M, Farrukh C, Mortara L, Thorn V (2009) Making the business case for technologies—a five step process guide. University of Cambridge, Institute for Manufacturing. ISBN 978-1-902546-81-0

# Chapter 11
# The Role of Policy Intervention for Technology Management: The Case of Swedish Hybrid-Electric Vehicle Technology

**Thomas Magnusson and Annika Rickne**

**Abstract** This chapter deals with how policy and strategic actions interact in the development of a technological field. As a company experiments with and evaluates various technological solutions, and continuously matches these with problems posed by specific market niches, the firms do this not as an isolated strategic activity, but in continuous interaction with numerous actors and processes. This implies that technology management has to be seen in a context of the working of the entire technological system and that technological innovation often needs policy interventions of various kinds. Our analysis illustrates how policy action comes into play in the emergence of a technological field, leading to implications for firms' technology management processes. We have chosen to focus our analysis on sustainable technologies in the auto sector, being a case where firms are dominant players, on a global market, but where they need to interact with market demands as well as with policy interventions. Our focus is on Sweden, where the auto sector is a dominant part of the national industry, and where sustainability issues have been placed in the forefront of policy debates.

## 11.1 Introduction

This chapter deals with how policy and strategic actions interact in the development of a technological field. Our starting point is that technological innovation is an evolutionary process, where experimentation leads to the development of a

T. Magnusson (✉)
Department of Management and Engineering, Linköping University, Linköping, Sweden
e-mail: thomas.magnusson@liu.se

A. Rickne
Institute for Innovation and Entrepreneurship, School of Business, Economics and Law, Gothenburg University, Vasastan, Sweden
e-mail: annika.rickne@handels.gu.se

D. Cetindamar et al. (eds.), *Strategic Planning Decisions in the High Tech Industry*, DOI: 10.1007/978-1-4471-4887-6_11,

range of alternatives, which are evaluated and selected in iterative processes involving a multitude of actors. Evolutionary economists speak about how variation, selection, and retention mechanisms form the basis for how to understand innovation, and point to the importance of scarce resources, uncertainty, and path dependency in deciding the outcome [1]. Such mechanisms become especially obvious in the evolution of new technological fields, where the level of experimentation is high and there is ambiguity both about performance levels of various technological alternatives and about market needs. Also, the very causality of development paths may be uncertain, where explicit customer demands interact with technology push and with guiding activities by policy actors.

In this setting, technology management becomes a highly uncertain and complex activity. As a company experiments with and evaluates various technological solutions, and continuously matches these with problems posed by specific market niches, the firms do this not as an isolated strategic activity but in continuous interaction with numerous actors and processes. This implies that technology management has to be seen in a context of the working of the entire technological system. Only by understanding the underlying motivations and the actions of all main actors in any specific innovation system can be a firm coherently manage its technological innovation.

One main type of actor influencing the emergence of technology and corresponding products and markets is governmental policy units. In fact, technological innovation often needs policy interventions of various kinds. This does not mean that governments or policy actors have a sole responsibility for setting in motion a favorable course of actions. On the contrary, the term 'governance' captures the fact that a range of actors and stakeholders can and perhaps should be involved, and that various arrangements for and processes of control are needed to advance societal objectives [2]. In fact, forms of suitable governance shift over time, sectors and countries. At some instances regulation, directives, or directed incentive schemes may be useful to achieve the desired goals; at others, public–private partnerships or open consortia may be better options. This implies that various stakeholders may take on different roles depending on the situation.

Based on this assumption, this chapter illustrates how policy action comes into play in the emergence of a technological field, leading to implications for firms' technology management processes. We have chosen to focus our analysis on sustainable technologies in the auto sector, being a case where firms (especially large, incumbent firms) are dominant players, on a global market, but where they need to interact with market demands as well as with policy interventions. Our focus is on Sweden, where the auto sector is a dominant part of the national industry, and where sustainability issues have been placed in the forefront of policy debates.

## 11.2 Our Approach

It is not always clear what constitutes effective governance or technology management to promote development and diffusion of innovations. Scholars adopting the technological innovation systems approach (TIS) point to a number of functions or *key processes* necessary for new technologies to develop and diffuse. The argument in this literature is that firms and policy makers alike need to be aware of these key processes and how well they function in the specific sector. Assessing the functionality of the innovation system along with these key processes helps reveal where progress could be induced and where there are bottlenecks to development [3, 4].

We will follow this approach and assesses how the functionality of the Swedish hybrid-electric vehicle technological innovation system evolved 1990–2010, identifying six key processes. For each of these key processes, the indicators of functionality were discussed in length in the underlying report by Magnusson [5].

1. *Knowledge development and diffusion*: The generation of breadth and depth of the knowledge base of the TIS, and the diffusion and combination of knowledge, taking into account different types of knowledge (e.g. scientific, applied, and patents) from different sources.
2. *Influence on direction of search*: The existence of incentives and/or pressures for actors to enter the TIS, and to direct their activities toward certain parts within the TIS (e.g. technologies, applications, or markets).
3. *Entrepreneurial experimentation*: The probing into new technologies and applications, unfolding a social learning process reducing uncertainty.
4. *Market formation*: The timing, size, and type of markets that have actually formed.
5. *Legitimation*: The social acceptance and compliance with relevant institutions: the new technology and its proponents need to be considered appropriate and desirable by relevant actors in different parts of the TIS to acquire political strength.
6. *Resource mobilization*: The mobilization of competence/human capital through education in specific scientific and technological fields as well as in entrepreneurship, management and finance, financial capital (seed and venture capital, diversifying firms, etc.).

Situated in a relatively small country with a limited domestic market, the Swedish vehicle industry depends heavily on exports. These exports are also a significant part of the total exports from Sweden. In 2008, exports of the Swedish vehicle industry amounted to 155 billion SEK, which is 13 % of the total Swedish exports [6]. Besides the Information and Communication Technology (ICT) sector, the vehicle industry is the largest sector in Sweden in terms of R&D spending [7]. The vehicle industry thus constitutes a very important part of the Swedish economy. The Swedish car manufacturers' production volume is small from a global perspective; in 2007, the Swedish car manufacturers' production amounted to 0.6 % of the total production worldwide. However, Swedish manufacturers of heavy vehicles (trucks and buses) have a much stronger position on the markets in which they are active. In 2007, the Swedish heavy vehicle manufacturers had a 28 % European market share [8].

The time period for the study is 1990–2010. The empirical study is based on studies of documents and loosely structured open-ended interviews with representatives for key industrial and public actors. Apart from this, to ensure a strong international component in the TIS analysis, the report also relies on interviews with representatives for leading international automotive manufacturers, key suppliers, and research funding agencies.

The technology in focus in the study is that underlying hybrid-electric automotive power trains. The automotive power train is the vehicular subsystem that generates and delivers motive power to the road surface. It constitutes a dominant part of the vehicle, incorporating some central parts such as engine, ignition system, transmission, and so on. It also determines some of the most important properties of the vehicle such as speed, acceleration, and traction power. In a traditional car, truck or bus, the internal combustion engine (ICE) is dimensioned according to the maximum power needs of the car during e.g., acceleration. This means that there is an excess capacity in normal driving and this excess capacity results in efficiency losses. However, if you combine the ICE with an electric motor to form a hybrid-electric power train, the engine can be downsized. As a result, you may gain efficiency. Moreover, being assisted by the electric motor, the internal combustion engine is allowed to consistently operate its most efficient range in terms of engine speed and torque. The electric motor can also be reversed and used as a generator to recover breaking energy, so-called regenerative breaking. Moreover, the hybrid-electric power train minimizes idling losses through automatic shutdown of the ICE, e.g., when waiting for green light at a crossing. Hence, the combination of the virtues of electric machines and internal combustion engines enables reduced fuel consumption, especially in driving conditions characterized by repeated starting/stopping and acceleration/deceleration. Combining electric propulsion with combustion engines may thus result in considerable gains in terms of fuel efficiency, compared to traditional vehicles.

## 11.3 Swedish Hybrid TIS Functionality 1990–2010

### *11.3.1 Knowledge Development and Diffusion*

Knowledge development and diffusion is about generation of breadth and depth of the knowledge base, and the diffusion and combination of knowledge, taking into account different types of knowledge from different sources. The hybrid power train comprises several interacting subsystems and components, which build upon a number of disparate fields of engineering knowledge. Requirements in terms of drivability, performance, fuel efficiency, safety, reliability and, not the least, cost put heavy restrictions onto the development. Hybrid-electric vehicle R&D is thus very much about systems integration. Vehicle manufacturers' abilities to operate as systems integrators hinge on a critical balance between system level architectural knowledge and component-specific knowledge [9]. A well-functioning hybrid electric vehicle TIS has to support R&D activities in order to develop this product-related technological knowledge.

Moreover, cost reduction in the vehicle industry to a large extent depends upon production volumes. Technological and organizational knowledge related to production is therefore an essential part of the required knowledge based in a well-functioning hybrid-electric vehicle TIS. Another relevant instance of knowledge is related to the interface between the vehicle and the market. Reference [10] describes this in terms of external product integrity, i.e., the fit among the function, structure and semantics of the vehicle and the customer's objectives, values, production system, lifestyle, use pattern, and self-identity. Consequently, one may discern three aspects of knowledge which are essential for a well-functioning hybrid-electric vehicle TIS. First, product-related technological knowledge (component and architectural knowledge) emanating from R&D activities where patents can be a useful indicator complemented with information on reported R&D activities and demonstrations of products through prototypes and field tests. Second, technological and organizational knowledge related to production, indicated by vehicles available for purchase and accumulated production volumes. Third, knowledge related to the interface between the vehicle and the market indicated by executed field trials with users and actual sales.

The functionality assessment thus starts by assessing product-related technological knowledge development, as measured by patents, prototypes, and field trials. Patenting records indicate that no specific car manufacturer had a particular lead in hybrid R&D until the mid-1990s. At this stage, Toyota and shortly thereafter Honda started patenting much more actively. This coincides with the start of these manufacturers' development efforts aiming for market introduction by the late 1990s. Just like other European car manufacturers, the Swedish vehicle manufacturers are largely absent in these records, signaling a relatively low degree of R&D activity by comparison to their Japanese contenders. This holds true also during the 2000s when the large American manufacturers Ford and GM (a few years later) started patenting more aggressively. Some Volvo activities can be discerned, but by comparison to their owner Ford, Volvo Cars' hybrid R&D appears marginal. According to the American database USPTO, Volvo Cars took four hybrid vehicle patents 1990–2008 (all of these were filed after 1999). This can be compared to 156 patents by Ford during the same period. Patenting records thus illustrate Ford's decision to centralize the group's hybrid technology R&D efforts after their acquisition of Volvo Cars. From a Swedish perspective, it is also worth noting that Saab Automobile does not appear at all in these patenting records.

On the international scene, Toyota and Honda took the lead during the early 2000s, pioneering the market introduction of hybrid cars. The significantly raised sales figures for Prius on the US market 2004–2008 served to reinforce Toyota's lead. Several additional car manufacturers introduced hybrid cars during the late 2000s, but by the end of the decade, no other car manufacturer came even close to Toyota in terms of accumulated production volumes and Honda appeared to have consolidated their position as an obvious runner-up in the hybrid segment. American and European car manufacturers (including the Swedish ones) were left behind. Their reluctance to enter the market prohibited learning from production and marketing of hybrid cars.

Being supported by first-tier suppliers such as Bosch and Valeo, European manufacturers instead emerged as front-runners in the implementation of micro-hybrid (start–stop) systems during the late 2000s. Volvo Cars was by no means first to enter the market as they introduced microhybrid systems in three of their Drive models 2008–2009. Still, with this fuel-efficient product line they appeared well in line with their European competitors and with a positive response from the market, sales, and production volumes have risen steeply since the market introduction. In this case, Volvo Cars benefits from the extensive experience of first-tier supplier Bosch, both in terms of product and production related knowledge.

The Swedish heavy vehicle manufacturers were active during the late 1990s, as well as in the mid and late 2000s, running several field tests and demonstrations of hybrid buses and trucks. These tests represent significant milestones for the Swedish hybrid-electric vehicle TIS in terms of development and diffusion of both product-related technological knowledge and knowledge related to the product-market interface. Amongst others, experiences from the tests in the late 1990s made Volvo finally abandon series-hybrid configurations in favor of parallel configurations. During interviews, respondents from Scania also emphasized the importance of field tests for their knowledge development, both in terms of technological possibilities and limitations, and in terms of learning how users perceive the new technology.

The heavy vehicle industry differs from the car industry in that no particular manufacturer has emerged as an obvious leader in hybrid R&D and production. While a few Japanese manufacturers have introduced products in the light-duty segment, several manufacturers (including the Swedish ones) aspire to take leading positions in the heavier segments. Judging from their market introduction of a hybrid bus in 2010, Volvo appeared to be well positioned in the international competition. With a product portfolio more focused on long-haulage trucks, hybrid technology may not seem as urgent for Scania, something that may explain a more restrictive strategy.

Consequently, information on reported R&D activities as well as demonstrations of functioning products through prototypes and field tests show that a certain level of product-related technological knowledge has been created and sustained within the Swedish hybrid-electric TIS 1990–2010. The varying intensity of reported R&D activities and demonstrations of functioning products naturally affected the level of knowledge generated in the system. In particular, one may discern a decline during the early 2000s and a recovery during the mid and late 2000s. Moreover, the heavy vehicle manufacturers' decisions to initiate product development aiming for market introduction during the late 2000s, and the consecutive field trials with customers also initiated some production and market-related knowledge generation in this sector. Still, by the end of 2010, apart from Volvo Cars' Drive micro hybrids, no significant volumes had been produced. This severely restricted the generation of production and market-related knowledge in the Swedish hybrid-electric TIS.

Our question is then how policy has influenced the development. Several sources underline the importance of the state funded R&D and demonstration programs for knowledge development and diffusion of hybrid vehicle technology in Sweden.

During an interview, a senior manager at Volvo explained that the financial support from the KFB electric and hybrid-electric vehicle demonstration program was a prerequisite to justify Volvo's hybrid heavy vehicle field trials in the late 1990s. He recalls that it was possible attain internal funding that would cover about half of the required budget. The other half had to come from external sources. Otherwise, Volvo's senior management would not have passed the project.

A general evaluation of the effects of state funded R&D programs directed at the Swedish automotive industry since the early 1990s provide additional support for the positive effects of these R&D programs [8]. The report specifically points at positive effects in a few areas, and hybrid-electric vehicle technology is one of these. According to the report, state funded R&D support was decisive for Volvo's ability to maintain hybrid technology knowledge when Ford acquired Volvo's car division in the late 1990s.

Referring to the state funded Green Car programs during the early and mid-2000s, a technology manager at Volvo Cars claimed that dedicated governmental funding in many cases was a prerequisite to justify long-term development efforts directed at fuel efficiency. At this stage (i.e. early and mid-2000s), it seemed as if demands on fuel efficiency were not sufficiently explicit, at least not in the customer segments that Volvo Cars aimed for. Under such circumstances, the only way of justifying R&D efforts addressing fuel efficiency was to have dedicated funding from external sources.

The establishment of the Swedish Hybrid Center in 2007, supported by the Swedish Energy Agency, signaled another step in the ambition in technological knowledge development, connecting academic researchers within relevant fields of knowledge to the vehicle manufacturers. While it is difficult at this stage to assess any results of this effort, this initiative may illustrate an insight that a deeper component-specific knowledge is necessary to build a sufficient level of integrative knowledge. This knowledge is required to move from a stage of limited field trials to the development of vehicles intended for sales on a commercial basis. The Swedish vehicle manufacturers do not possess sufficient resources to develop this knowledge individually all by themselves. Collaboration between vehicle manufacturers and academic research is put forward as a means to overcome this problem.

Consequently, it seems as if state funded R&D and demonstration programs filled a role to develop and maintain a certain level of knowledge in the Swedish hybrid-electric vehicle TIS. In particular, this has been valuable to mobilize engineering resources during periods when the automotive manufacturers experienced relatively weak motives to invest in hybrid R&D and when it was difficult for them to justify such efforts internally. Hence, these programs occasionally served to compensate for a low legitimacy in the innovation system. However, this only applies to certain aspects of knowledge development and diffusion. The state funded R&D and demonstration programs were important to support pre-commercial technology development. They also supported field trials with users, particularly in the case of heavy vehicles. This has helped vehicle manufacturers developing product-related technological knowledge. It has also helped heavy vehicle manufacturers to understand user requirements and behavior, i.e., knowledge related to the product-market interface. Altogether, this has reduced their take-off distance for initiation of product development projects directed at commercialization. However, while the KFB program during the 1990s dedicated a share of the total budget to social science research, addressing issues such as market developments for electric and hybrid-electric vehicles and the role of such vehicles in the society, R&D funding programs during the 2000s have been largely techno centric.

### 11.3.2 Influence on Direction of Search

Influence on direction of search is about incentives and/or pressures for actors to enter the TIS, and to direct their activities toward certain parts within the TIS (e.g. technologies, applications, or markets). The automotive industry is very public, attracting a lot of interest from media. Each year vehicle manufacturers worldwide present their visions about the future. These are often materialized in concept vehicles, which are displayed at auto shows, which is one indicator of a vehicle manufacturer's long-term direction in terms of technology and product strategy, combined with manufacturers' public announcements.

Responding to raised demands on low-emission vehicles, Volvo's display of their Environmental Concept Car ECC in 1992 represents an early milestone in terms of Influence on direction of search. By comparison to a traditional ICE-based power train, Volvo's microturbine-based series-hybrid offered clear advantages in terms of reduced emissions of Nitric Oxides (NOx), Particulate Matter (PM), and Hydrocarbons (HC). These substances are directly related to urban air-quality problems, which were a very important issue for vehicle manufacturers during the early 1990s. The positive reactions that Volvo got encouraged them to engage in further development of their microturbine-based series hybrid, developing heavy concept vehicles (a truck and a bus) based on similar technological concepts in the mid-1990s, and also to proceed to a stage of field trials. However, it is important to note that these vehicles never were intended for production. By the late 1990s, Volvo abandoned the microturbine-based series hybrid. This decision can partly be explained by lessons gained from field trials. However, the weakened Californian ZEV-regulation as well as incremental improvements of traditional engine technologies and emission reductions through widespread implementation of catalytic converters [11, 12] also contributed. Considering these changes, a radical shift to alternative power train technologies for the purpose of reducing NOx, PM and HC emissions, appeared less plausible during the late 1990s. Instead, $CO_2$-emissions and fuel efficiency started to emerge as the prime motive for hybrid-electric vehicle technology, turning the focus from series hybrids to parallel or power-split hybrid configurations.

As indicated by patenting numbers, Toyota's market launch of the hybrid car in Prius in 1997 resulted in a lot of R&D activities among competitors in the area of hybrid vehicle technology. Volvo Cars displayed a hybrid concept car in 1999, comprising a similar hybrid configuration as Toyota Prius, i.e. a power-split hybrid.

Due to Ford's acquisition of Volvo Cars in 1999 and Volvo's subsequent attempt to acquire Scania, this was a period of great turbulence for the Swedish vehicle industry. Moreover, with a relatively stable fuel price, customers were not expected to be prepared to pay for expensive fuel saving hybrid technologies. By consequence, incentives to further advance hybrid technology toward commercialization were limited and the functionality in terms of Influence on direction of search in the Swedish hybrid-electric vehicle TIS appear relatively weak during the early 2000s.

This period was broken in the mid-2000s, as crude oil prices started to soar. In particular, this affected the hitherto relatively low US fuel prices. Heated public debates on global warming and $CO_2$ emissions served to reinforce the drive for fuel efficiency and 2004–2007, there was a 5-fold increase of Toyota Prius sales on the American market. In 2006, Volvo's CEO announced intentions launch a hybrid truck 3 years later and he also announced that Volvo intended to launch hybrid buses and wheel loaders. During interviews, Volvo managers directly referred to Toyota as their role model in this endeavor, claiming an ambition to become a corresponding hybrid technology leader in the heavy vehicle segment. Volvo's announcement also seems to have affected Scania. During an interview, a senior manager at Scania expressed his admiration for the bold statement from Volvo's CEO. Moreover, Scania followed suit displaying a series-hybrid concept at the UITP Mobility & City Transport Exhibition in 2007, and they subsequently transferred this hybrid development project to their R&D centre in Södertälje, aiming for market introduction by 2012.

While financial crisis and reduced fuel prices during the late 2000s reduced the pace of development, the long-term direction and aim for significantly increased fuel efficiency appeared set for the heavy vehicle industry. This also applies to the car industry. Saab Automobiles display of hybrid concept vehicles in the late 2000s, as well as Volvo Cars' display of the plug-in hybrid concept ReCharge in 2007 can be interpreted as signs of aspiration to take part in the race toward increased fuel efficiency. Volvo Cars' introduction of the Drive model range in 2009, comprising three models equipped with micro-hybrid systems as well as their announcement that they aim for production of plug-in hybrids by 2012 (albeit with a reservations that a commercialization will rely on subsidies) also indicate long-term direction in terms of hybrid technology, aiming for leaps in fuel efficiency, and reduced $CO_2$ emissions.

As regards the influence of policy, the Californian ZEV-mandate obviously promoted electrification and—especially in the case of Volvo—the development of hybrid-electric vehicles. Volvo considered the microturbine based series-hybrid as a promising means to reduce emissions of local pollutants and fulfill the proposed ZEV regulation during the 1990s. However, this law-enforced market did not materialize. Instead, fuel efficiency emerged as the main driver for hybrid-electric vehicle technology. Toyota introduced their Prius in the late 1990s and the global scene saw an emerging debate on $CO_2$ emissions and global warming. However, during this period, the vehicle manufacturers had relatively weak incentives to pursue this development. Fuel prices were low and the voluntary agreements between the EU commission and European car manufacturers on average $CO_2$ emissions were not forceful enough to promote further commercialization.

It was not until the mid-2000s that the motives were sufficiently strong for the Swedish vehicle manufacturers to advance hybrid technology further, engaging in product development directed at commercialization. At this stage, the manufacturers felt strong signals from several sources. First, the sales figures of Toyota Prius on the US market showed that there actually was a market for hybrid vehicles. Second, souring fuel prices made it possible, or at least plausible, to justify hybrid vehicles based on cost rationales. Third, a specific governance arrangement that Scania managers repeatedly have referred to during interviews is a contest between bus manufacturers to deliver fuel efficient and low polluting transport solutions to Transport for London. According to these managers, this was a prime reason why Scania planned for market launch of their series hybrid bus by 2012, plans which they later canceled due to discouraging results from their field tests in Stockholm. The Transport for London initiative was even more important for Volvo's development, as this offered a possibility for extensive validation of their hybrid bus. This governance arrangement was also associated with financial support for investments in fuel-efficient buses, something that served as an indication of an upcoming bridging market for hybrid buses.

In addition to this, the more forceful policy on transport-related $CO_2$-emissions adopted by the EU commission in 2007 comprised clear limits on average emissions from new cars as well as plans for stepwise reduction of these limits. This presented strong signals to the European car manufacturers, providing clear targets to aim for. Subsidies and tax reductions associated with $CO_2$-emission threshold levels in individual EU-countries were highly influential as well. For example, the 15 % of governmental subsidy for the purchase of cars with $CO_2$-emissions below 105 g/km in Belgium provided an obvious target for the engineers at Volvo Cars in the development of the smallest Drive models, and thus had an important influence on the decision to implement micro-hybrid technology in these models.

### 11.3.3 Entrepreneurial Experimentation

Experimentation and probing into different technologies and applications is necessary to reduce uncertainty and facilitate learning in a technological innovation system [4]. For its assessment we rely on two main indicators: the variety in technological configurations under development, and the number of different product applications in which Swedish vehicle manufacturers have implemented hybrid-electric power trains.

The entrepreneurial experimentation differs between the Swedish vehicle manufacturers. In terms of the variety of hybrid configurations under development, Volvo's activities seem to have peaked during the late 1990s. Encouraged by positive feedback from their display of the microturbine-based series-hybrid concept car ECC during the early 1990s, they extended this concept to two other applications, a concept truck and a concept bus. Volvo also took the microturbine-based series-hybrid one step further, testing it in an inner-city bus application in regular operation. Concurrently, they conducted field tests with an ICE-based series-hybrid distribution truck. Volvo's car division was also very active during the late 1990s. They developed and tested several different ICE-based hybrids, including five versions of a series hybrid as well as a mild hybrid configuration. In 1999, they also displayed a concept car with a power-split hybrid [13].

The level of entrepreneurial experimentation at Volvo decreased significantly during the early 2000s. This was to a large extent because of Ford's acquisition of Volvo Cars and their subsequent decision to concentrate the group's hybrid R&D efforts to USA. During this period, the R&D efforts also started to converge toward an ICE-based parallel full hybrid configuration, which Volvo primarily tested and demonstrated in a distribution truck application. Their Construction Equipment division also conducted pre-commercial R&D, testing an ICE-based parallel hybrid configuration in a wheel loader application. Moreover, initiated by Renova's efforts and interest, refuse trucks emerged as a viable application for hybrid-electric vehicle technology.

The decision to initiate product development directed at commercialization further enhanced the convergence in terms of hybrid configurations at Volvo. A common platform strategy would mean that all relevant product divisions could share the same technology. This was deemed necessary to reach sufficiently high production volumes, which were crucial to justify the investment. To realize this, Volvo managers selected the most versatile hybrid configuration and positioned the development project at Volvo Power train, an organizational unit that operates across the product divisions. The hybrid platform strategy addressed several product applications such as distribution trucks, refuse trucks, inner-city buses, and Penta (industrial engines/boat engines). In the Penta application, the hybrid power train was primarily intended as an integrated auxiliary/back-up electric power unit, comprising both engine and power generator. Hence, while the degree of entrepreneurial experimentation reduced in terms of technological variety, it increased in terms of variety in product applications.

In 2004, Ford released a hybrid version of their Escape SUV on the American market. It was equipped with a full power-split hybrid. At this stage Ford had intentions to expand their hybrid model range and introduce hybrid technology on the European market. Following these intentions, Ford's management acknowledged the virtue of having R&D capabilities close to the market and reconsidered their decision to concentrate the group's hybrid technology R&D to USA. Thus, the Ford management assigned Volvo Cars' Torslanda plant as their European hybrid technology center. However, Ford ran into severe financial problems by the mid-2000s, which slowed down the plans to expand the hybrid model range. As Ford announced plans to sell of Volvo Cars by the late 2000s they also withdrew the hybrid R&D funds. However, Volvo Cars still maintained activity in the hybrid technology field, developing, and initiating field tests of a plug-in hybrid car with a diesel engine. Moreover, they expanded the range of hybrid technology, developing, and launching three car models equipped with micro-hybrid systems.

Scania conducted some R&D on hybrid trucks, especially during the mid and late 2000s, using a parallel hybrid configuration. However, Scania particularly focused their hybrid R&D on bus applications. This is natural since the heavy truck models

Scania manufactures primarily are intended for long-haulage applications. According to [14], Scania has tested and demonstrated around 30 buses since the 1980s with different hybrid power trains. This includes buses equipped with both parallel hybrid and series hybrid configurations. Scania often built these buses in collaboration with the Danish body manufacturer DAB, a company that Scania owned 1995–2002. Their focus on bus applications explains why they maintained an interest for the series-hybrid configuration, even though the automotive industry in general turned its main focus toward parallel or power-split hybrids.

Analyzing the influence of governance on entrepreneurial experimentation will result in similar results as the analysis of the influence of governance on Knowledge development and diffusion. State funded R&D and demonstration programs have made it possible for the Swedish vehicle manufacturers to test various hybrid configurations in a number of different product applications.

A retrospective evaluation report of the Swedish KFB program for research, development, and demonstration of electric and hybrid-electric vehicles 1993–2000 argues that the program could have had a stronger focus on hybrid technology for heavy vehicles, rather than the present focus on electric cars [15]. Due to the American ownership of the Swedish car manufacturers, it would probably have been easier to influence the degree of activity at the heavy vehicle manufacturers than at the car manufacturers. However, it is still questionable to what extent the heavy vehicle manufacturers would have been able to justify a further expansion of their R&D, testing and demonstration activities at this stage. Such decisions are still determined through strategic considerations. An important lesson from the KFB program was that such a program has to involve the vehicle industry more actively, engaging the car manufacturers in a dialog both in the planning and execution of the program. These lessons were transferred to the subsequent Green Car programs, which were based on a common agreement between the Swedish state and the Swedish vehicle manufacturers.

While the Californian ZEV-mandate, the implementation of environmental zones, and the generally strong focus on local air emissions during the early 1990s particularly encouraged the development of electric cars, the turned focus toward fuel efficiency and $CO_2$-emissions in the late 1990 and 2000s resulted in a larger variety in terms of hybrid configurations. This includes series, parallel, and power-split hybrids as well as micro, mild, full, and plug-in hybrids. An important note, however, is that in terms of product applications recent market-based incentives, such as subsidies, tax reductions, and regulation on $CO_2$-emissions have primarily been directed at cars rather than heavy vehicles. Hence, these governance mechanisms have not been very effective to increase variety in product applications. However, heavy vehicles are more exposed to local regulations, such as environmental zones that restrict operation of diesel-engine powered vehicles in inner city areas order to improve air-quality and reduce noise. Such regulations have indeed encouraged experimentation of hybrid technologies (and other alternative power train technologies and fuels) in heavy vehicle applications such as distribution trucks, inner-city buses, and refuse trucks.

## *11.3.4 Market Formation*

New technologies are often relatively crude and perform badly according to an established set of performance parameters. This means that they are rarely competitive on regular markets, which are dominated by the existing technologies. These technologies have been subject to years of continuous improvement and adjustment to fit the exact needs and requirements of users, and have been standardized and production processes have been developed and fine-tuned, resulting in reduced manufacturing costs. Therefore, so-called "nursing markets" are often necessary to support an early diffusion of the new technology onto the market. Nursing markets constitute protected spaces, which allow for testing, experimentation, demonstration, adaptation, and further improvement [16]. Such markets can also provide a prosperous interaction between manufacturers and so-called "lead-users" [17], i.e., professional users who tend to push technology development by defining requirements on new technology, by evaluating new product designs and by taking initiatives in the development of new product concepts and ideas.

The next step in a TIS market formation is the emergence of bridging markets [18]. Such markets allow for production volumes to increase and may thus provide a prerequisite for the opening of a mass market for the new technology. Bridging markets are niches or segments of the market, such as particular customer groups, geographical areas, or domains of application. In these segments, the particular merits of the new technology are of great value, while the deficiencies are acceptable. Hence, such bridging markets provide a necessary basis for volume increases that support production learning and incremental innovation. In turn, this will enable improvements of the price/performance ratio of the new technology, eventually making it competitive with established technologies on mass markets.

The assessment of the market formation will thus focus on the stage of market development, including the availability of nursing markets, as well as bridging markets that may open up a mass market for hybrid vehicle technology. In the assessment of market formation, it is important to make a distinction between the market for hybrid cars and the market for heavy hybrid vehicles. Buying behaviors differ markedly between cars and heavy vehicles. These two markets have also reached different stages in terms of market development.

In the hybrid car market, Toyotas careful entry with limited sales on the domestic market in the late 1990s can be interpreted as an attempt to address a local nursing market. The limited sales reduced the level of risk and the geographical proximity made it easier for Toyota's organization to respond to signals from the field and to correct potential problems with the new technology. Lessons gained from the field were fed back to product development, resulting in the release of an upgraded, and improved Prius 2 years later, a vehicle intended for export markets. From a Swedish perspective, such nursing markets for hybrid cars did not emerge during the late 1990s. According to [15], this was essentially because no hybrid vehicles were available on the Swedish market at this stage.

The subsequent development of the hybrid car sales in the US during the 2000s can be interpreted as a bridging market for hybrid vehicle technology. After a period of relatively modest sales in the early 2000s, hybrid sales took off remarkably during the late 2000s, propelled by soaring fuel prices and a heated global warming debate. The hybrid car's particular merits in terms of its symbolic values, made an emerging segment of environmentally concerned American car buyers embrace the new technology.

In terms of expressing these symbolic values, the second generation Toyota Prius was obviously superior among the cars available on the US market. Compared to the hybrid versions of Honda's standard models Civic and Accord, the dedicated hybrid brand made it was easier to distinguish a Toyota Prius. This was important for hybrid owners who wanted to make a statement and convey a message with their car purchase. The distinct body design of second generation Prius reinforced this feature. Moreover, the Prius mid-size body could attract a broader market segment than Honda's peculiar two-seated Insight model, which "had little appeal for mainstream customers" [19, p. 169]. The release of several hybrid versions of standard car models from a number of car manufacturers (not the least Toyota) by the late 2000s indicate that the vehicle manufacturers may have foreseen that hybrid vehicle technology was about to take the step from the bridging market to the mass market. But even though several other hybrid car models and manufacturers were available on the market, Toyota Prius still dominated the hybrid car market by the end of 2009. This indicates that hybrid vehicle technology had not yet taken the step from a stage of bridging market to a mass-market stage.

While both Toyota and Honda introduced hybrid cars in Europe in the early 2000s, sales on the European market were more limited. By contrast, Europe has been the prime target market for micro-hybrid technology. Volkswagen implemented a start–stop system already during the 1990s in a version of their Golf model called "Ecomatic" and in the most fuel-efficient versions of the microcar Lupo and Audi A2. However, the market reception was cold at this stage. Volkswagen, therefore, decided to cease the production after a few years. The market failure was explained by the discomfort that car drivers felt when the engine automatically shut off at red lights. However, by the late 2000s the technology experienced a revival. Several major European car manufacturers implemented micro-hybrid systems to supplement their most fuel-efficient engine alternatives, reducing fuel consumption even further. First-tier suppliers Bosch and Valeo expect a rapid market growth for these systems, more rapid than for full hybrids. Hence, after its re-entry in the late 2000s, micro-hybrid technology has moved rapidly toward a stage of mass market.

The market for heavy hybrid vehicles is more diverse than the hybrid car market. Particularly in the US, there is a limited market segment for hybrid buses, mainly driven by local regulations that restrict operation of traditional diesel engine powered buses in areas which are particularly sensitive to air pollution and noise. This market is served by specialized firms that equip buses or bus chassis with hybrid power trains. Moreover, a market for light-duty trucks has recently begun to emerge in Japan, primarily served by Japanese manufacturers such as

Toyota-owned Hino. The Swedish heavy vehicle manufacturers, however, have not been active in this segment.

The Swedish heavy vehicle manufactures have repeatedly benefitted from field trials in collaboration with local customers. Volvo has benefitted from collaboration with the distribution company TGM in the field trials of their series hybrid truck in the late 1990s and the refuse and recycling companies Renova and Ragn-Sells in the case of the parallel hybrid in the late 2000s. Renova also took the role of a lead user, promoting the development of hybrid-electric vehicle technology during the early 2000s by initiating development and testing of refuse trucks with electric auxiliary equipment. Two of the most important customers forming local nursing markets for the Swedish heavy vehicle manufacturers have been the municipal bus operators SL (Stockholm) and Göteborg Spårvägar. Close collaborations between Scania and SL and between Volvo and Göteborg Spårvägar have been instrumental to test the field performance of the hybrid technology and feed user experiences from the new technology back to R&D.

Volvo's decision to initiate development of heavy hybrid vehicles aiming for market launch indicates a belief that bridging markets will emerge and that hybrid technology eventually will take the step to a mass market. In this effort, Volvo has also proceeded beyond restricted collaboration with local partners, delivering field trial vehicles to customers in e.g., Great Britain and France. In particular, Transport for London's extensive test and evaluation program was quite important for Volvo's validation and further commercialization of their hybrid bus. During the second half of the 2000s, however, the market for heavy hybrid vehicles was still in a nursing stage and the financial crisis meant that the imminent market development for hybrid technology was associated with question marks, at least on the short term.

As regards the influence of policy, apart from supporting knowledge development, the Swedish KFB program for research, development, and demonstration of electric and hybrid-electric vehicles 1993–2000 had an objective of influencing market development. However during its initial years, the program experienced severe problems in obtaining cars. The program directors had overestimated the vehicle manufacturers' capabilities as well as their actual interest in supplying cars. According to an evaluation report [15], this particularly relates to the capabilities and interest of the Swedish manufacturers. Saab Automobile ceased their activities in electric and hybrid cars as they were acquired by GM and Volvo developed hybrid-electric cars during the 1990s, but this project was near closure in 1996/1997. By consequence, the program had to purchase vehicles from large international manufacturers (primarily the French manufacturers Renault and PSA).

The KFB program and the parallel public procurement program orchestrated by Nutek provided financial funds and technical support for municipalities and other local market actors to purchase battery electric and hybrid cars. The idea was to push the development forward by creating a demand for these cars. These coalitions of buyers were supposed to be strong enough to enforce challenging product requirements onto the car manufacturers. Hence, the KFB program and the parallel Nutek program aimed at creating bridging markets for electric and hybrid-electric vehicles. However, the evaluation report states that the program in the end had a small impact on market development. It also concludes that in future programs, the vehicle distribution network (i.e. the car retailers) has to be involved to ensure availability of vehicles, as well as spare parts, service, and maintenance [15]. In the end, these actors were not sufficiently powerful. Furthermore, it appears that the vehicle manufacturers and other actors in the vehicle supply chain were not prepared to serve a bridging market for hybrid-electric cars at this stage.

Even though the heavy vehicle part of the KFB program was significantly smaller than the car part, this part of the program probably had a greater impact on technological development. It provided a basis for a constructive interplay between vehicle manufacturers and users. However, the collaboration was occasionally hampered by diverging expectations. In 1999, Göteborg Spårvägar purchased two microturbine-based series hybrid buses from Volvo for operation in regular traffic. In an interview, a Volvo manager argued that the decision to operate the buses in regular traffic at this stage was premature. The operation meant that the buses were directly compared with traditional buses and the new technology was thus exposed to too high demands in terms of reliability and availability. There was not a sufficient acceptance for the downtime, which was significantly higher for the hybrid buses than for traditional buses. Hence, also in this case, the diverging expectations can be interpreted as caused by differing perceptions about the stage of market development and the maturity of the technology.

The voluntary agreements between the EU and the European car industry on $CO_2$-emissions did not have any substantial effect on the market development for hybrid vehicle technology. By contrast, the stricter EU-regulation proposed by the late 2000s, combined with tax incentives and subsidies at certain $CO_2$-emission threshold levels had a stronger effect. These were implemented in several EU countries during the late 2000s. Apart from providing a direction for R&D activities, these emission thresholds provided the industry with valuable instruments for product proliferation. Figures such as 120 or 105 g/km, associated with tax incentives or subsidies, simplified market communication, providing a useful tool for the car manufacturers in the promotion of their most fuel-efficient car models.

### *11.3.5 Legitimation*

Legitimation is about social acceptance and compliance. A new technology needs to be considered appropriate and desirable by relevant actors in the innovation system to acquire political strength. While a number of different actors are involved in the Swedish hybrid-electric vehicle TIS, the vehicle manufacturers and key subsystem suppliers are arguably key actors. In order to advance further, hybrid technology therefore has to be considered appropriate and desirable within these organizations.

The global automotive industry is marked by production overcapacity and shrinking margins. Most manufacturers balance on the edge of survival. By consequence, failed ventures in innovative R&D could have severe consequences and potentially even jeopardize the complete business. Hybrid development projects thus have to present strong motives to attain organizational support. The assessment of TIS functionality according to the legitimation function will thus focus on the strengths of the motives to conduct hybrid R&D and the ability to justify investments related to hybrid technology. In particular, the assessment will focus on the ability to justify the step from a stage of concept or prototype development to the development of vehicles intended for production and sales on a commercial basis. This is because the level of the investment (and thus the risk level) associated with a commercial venture is by an order of magnitude higher.

The limited size of the Swedish automotive industry means that it cannot support a complete supplier network. Nevertheless, Toyota's way of organizing hybrid vehicle R&D and manufacturing indicates that proximity between vehicle manufacturers and suppliers of key subsystems is advantageous. Therefore, the existence of committed suppliers of key subsystems to the hybrid system is also a useful indicator of legitimation in the Swedish hybrid vehicle technology innovation system.

The prime motives to conduct hybrid technology R&D changed substantially during the time period 1990–2010. While the focus during the 1990s primarily was set on local air pollutants, it gradually turned toward fuel efficiency and $CO_2$-emissions. However, neither local emissions nor the emerging global warming debate during the late 1990s provided sufficient motives for the Swedish vehicle manufacturers to engage further than development of concept vehicles and prototypes. Still during the early 2000s, motives for commercial product development appeared relatively weak. There are at least three reasons for this. These reasons are both related to external factors and to organizational routines and mindsets.

First, within these organizations a number of different project proposals compete to achieve internal R&D funding. In most cases, development projects that address explicit customer demands win this internal competition and achieve funding. If customers clearly state that they are prepared to pay for certain attributes, the perceived risk of engaging in such a development project is low. During the early 2000s, customers did not pose any explicit demands of the potential benefits associated with hybrid vehicle technology, and therefore these project proposals attained limited organizational support. Second, the additional cost associated with the purchase of a hybrid vehicle can be justified by reduced fuel expenses while using the vehicle. However, the fuel price was relatively stable during this period and it was difficult to justify the purchase of hybrid vehicles based on such cost/benefit estimates. Third, as investments in product development were assessed on a product level, it was difficult to justify the additional cost associated with hybrid technology development. The product-based calculation model simply did not fit with long-term hybrid technology investments. To justify these investments, a longer time frame was required and the calculation had to assume a sharing of learning costs between several products and product generations. The fact that an unusual top-down decision eventually (in 2006) was needed at Volvo to initiate the development of heavy hybrid vehicles for commercial production and sales show that existing decision-making routines within the organization were insufficient for such a bold decision.

During the late 2000s, several heavy vehicle manufacturers presented hybrid concept vehicles and announced plans to launch such vehicles on the market. At this stage, a number of converging factors served to reinforce the legitimacy of hybrid technology, making the decision to initiate product development directed at commercialization more viable. The crude oil price had started to climb dramatically and this altered the cost/benefit estimates. It was still difficult to justify the purchase of a heavy hybrid vehicle based on the current fuel price, but it was possible (and appeared reasonable) to base estimates on projections of a long-term trend of significantly raised fuel prices. Under such circumstances reduced fuel costs would make it rational for customers to prefer the more expensive heavy hybrid vehicles. Hybrid technology could thus be justified based on cost rationales. Moreover, the intense debate on global warming and the emerging global consciousness on the sense in and necessity of drastically reduced $CO_2$-emissions provided additional arguments in favor of fuel saving technologies. The sales figures of Toyota Prius on the US market also showed that it was possible to make business out of hybrid vehicle technology. Thus, it was tempting for heavy vehicle manufacturers to repeat Toyota's success in the heavy vehicle segment. The focus on fuel efficiency, global warming, and $CO_2$-emissions became particularly strong by the end of the time period. However, at this stage the industry was also seriously affected by financial crises and a severe downturn of the market.

Due to a generally lower mileage, it may have been even more difficult to justify hybrid technology using rational cost/benefit estimates in the case of cars. However, car buyers found other arguments to adopt hybrid technology, especially on the US market. According to [20], there is a lot of symbolic value associated with fuel efficiency in general and with the purchase of a hybrid car in particular. In interviews with American hybrid car owners, they found that their car purchases rarely were based on any rational economic calculations. Instead, "buyers of HEVs talked about making a commitment" and "for several hybrid buyers the idea of commitment included setting an example, being a pioneer" (p. 1221). Hence, it is important for these buyers that it actually shows that they are driving a hybrid car.

Several market launches of new hybrid car models from Japanese and American manufacturers show that hybrid technology gained legitimacy in the car industry during the mid-2000s. Still the European manufacturers (including the Swedish ones) appeared hesitant to introduce full hybrids, preferring the less radical (and less expensive) option of complementing traditional power trains with micro-hybrid systems. There may be several reasons for this, but comparing US and European market characteristics, one can distinguish three particular factors, which may have affected the European manufacturers' product strategies. The first relates to the volatility of fuel costs. Although European prices still were higher than American, fuel prices in Europe have increased on a relatively continuous basis. By contrast, US fuel prices increased dramatically during a short period in time, 2004–2008 (first half). This urged car buyers to search for alternatives. Second, the diesel car is a strong competitor on the European market, presenting a readily available and lower priced option for consumers interested in fuel efficiency. European car manufacturers also responded to customer demands on fuel efficiency by introducing their most fuel-efficient models (typically propelled by small direct-injection diesel engines complemented with micro-hybrid systems to further reduce fuel consumption) under special brands. Third, while a distinct and growing niche of environmentally concerned car buyers embraced the hybrid car Toyota Prius on the US market; such a distinct and significantly growing niche did not arise in Europe or at least it did not adopt Toyota Prius as its defining artefact.

As for suppliers of key hybrid subsystems, the large Swedish-Swiss electric equipment manufacturer ABB was deeply involved in hybrid-electric vehicle R&D during the 1990s, developing electrical machines and power electronics for Volvo's concept, prototype, and demonstration vehicles. For this reason, ABB formed a particular division dedicated to hybrid vehicle technology, ABB Hybrid Systems, in the mid-1990s. However, by the turn of the century their senior management decided to terminate the company's efforts in hybrid vehicle technology. This decision was taken in connection with the severe corporate financial crisis that ABB suffered in the late 1990s. As a result ABB initiated a through restructuring, selling off a number of business areas and focusing on their core businesses in electric power transmission and distribution, as well as industrial automation. At this stage, ABB had invested some 80–100 MSEK in hybrid-electric vehicle technology over a period of 10 years [21, p. 381]. Volvo Cars had recently terminated the series hybrid development efforts that ABB had been part of. Moreover, BMW had recently (in 1999) terminated an electric vehicle development project for which ABB had supplied electric drive systems. There were thus serious doubts on the future business potential of electric and hybrid-electric vehicle technology. Moreover, ABB managers realized that the high manufacturing volumes and low profit margins in the automotive industry did not correspond very well with their traditional line of business. The vehicle manufacturers demanded a detailed insight and a high degree of control into their suppliers' operations and the ABB managers were not prepared to accept such a business proposal. Hence, they took a strategic decision that automotive technology should not be part of their business and that they should not serve as a supplier to the vehicle industry in the future.

Related to the influence of policy, R&D investments in the vehicle industry tend to need justification in terms of explicit or perceived market requirements. Strict regulation could be used for this purpose. The law enforced requirements related to local air pollutants and demands on zero and low-emission vehicles during the 1990s were sufficient to justify the development of hybrid concept vehicles and prototypes. However, the requirements were weakened and traditional ICE-powered vehicles became cleaner by the end of the 1990s. In the end, the regulatory requirements did not provide sufficient motives for the Swedish vehicle manufacturers to engage further in hybrid R&D.

Volvo Cars' redirection of the product strategy toward fuel efficiency in the late 2000s can be ascribed to a sudden insight that fuel efficiency had become an important competitive parameter on the market. These customer requirements on fuel efficiency were supported by governance arrangements in terms of $CO_2$ declarations and threshold levels that enabled benchmarks between different car models. At this stage, Volvo Cars lagged behind several of their European competitors. Product development was urgent to catch up with the competitors' lead.

The quick development and introduction of the new fuel-efficient Drive model range, with micro-hybrid systems in three of these models, was an important part of Volvo's strategy to catch up.

For the heavy vehicle manufacturers, soaring oil prices during the mid-2000s instead provided decisive motives to engage further and initiate R&D directed at commercialization. By extrapolating these trends, it was now possible to justify commercialization of fuel saving technologies. Governmental subsidies could help stimulate such demand further. However, a project manager at Volvo Power train underlined that a strategic decision to introduce a hybrid vehicle on the market cannot rely on governmental support. It is important that the development project is commercially feasible without subsidies. In some cases, governmental subsidies can even be counterproductive in terms of legitimation. An interview respondent with long experience from R&D on electric machines and with positions both within industry and academia projected that ABB eventually would reconsider their decision and re-enter as a supplier of electric machines to hybrid electric vehicles. In particular, the heavy vehicle industry should be interesting for ABB since the size of the electric machines in these applications correspond relatively well with the traction machines that ABB currently produce. What is more, heavy vehicle production volumes are much lower than car production volumes. However, a prerequisite for such a strategic re-entry would be that the vehicle manufacturers show that they are serious about hybrid technology. One way of showing this is to make comprehensive investments in the new technology. According to the respondent, the seriousness of the vehicle manufacturers' efforts in hybrid vehicle technology can be questioned as long as a significant portion of their R&D is funded through governmental R&D subsidies.

### 11.3.6 Resource Mobilization

A vehicle consists of thousands of components. In order to be approved for production, each component has to fulfill detailed requirements. Furthermore, the complete vehicle has to be validated to ensure strict system level requirements in terms of safety, manufacturability, reliability, durability, cost, drivability, comfort, serviceability, and so on. Vehicle development projects thus comprise very elaborate validation activities. By consequence, such projects are extremely engineering intensive endeavors.

The hybrid-electric power train builds upon integration of different fields of engineering knowledge. Some of these fields are very familiar to the vehicle industry. Vehicle manufacturers have since long built a competitive edge based on expertize in areas such as mechanical engineering and internal combustion engines. However, the vehicle manufacturers are less familiar with other fields of engineering knowledge, such as electric propulsion systems and battery technology.

The assessment of the resource mobilization function of the Swedish hybrid-electric vehicle TIS will focus on the number of available engineers with competence in relevant fields, in relation to the experienced need as expressed by key actors in the system. The functionality assessment will also incorporate a discussion on how the Swedish vehicle manufacturers have addressed the problem of mobilizing the necessary engineering resources for their hybrid-electric vehicle development efforts.

In general, the hybrid vehicle concept, prototype, and demonstration projects during the 1990s and could be conducted by small dedicated teams of engineers. Therefore, the need for engineering resources was relatively limited in the Swedish hybrid-electric vehicle TIS until the mid-2000s. At this stage, the need for engineering resources escalated during a short period of time. This was primarily due to Volvo's decision to initiate the development of heavy hybrid vehicles in 2006 aiming for market introduction 3 years later, as well as Scania's decision to transfer their hybrid bus development to Södertälje, initiating extensive field tests and aiming for market introduction by 2012. Moreover, at this stage, Ford decided to locate their European hybrid R&D at Volvo Cars. All these vehicle manufacturers competed for similar engineering resources and by comparison to the previous concept, prototype, and demonstration projects, the engineering input required for the development of vehicles aiming for market introduction was by an order of magnitude higher. Hundreds of engineering specialists were required in order to validate the vehicle for production. To fulfill their immediate need for engineering resources with relevant competencies, the vehicle manufacturers had to devise a number of different approaches, such as external and internal recruitment as well as engaging consultants. A group manager at Volvo Power train explained that it was particularly difficult to find engineers who had knowledge within the "new" fields (e.g. electric machinery, energy storage, and power electronics) in addition to experience from vehicle development and production. Although subsystems such as batteries and electric drives are sourced from external suppliers, this combination of engineering knowledge and industrial experience was required to provide the suppliers with accurate specifications in terms of product quality, terms delivery, product support and so on. Moreover, it was required to be able integrate the hybrid power train as a system.

During the time period 1990–2010, the Swedish vehicle manufacturers used a number of long-term strategies to acquire relevant engineering knowledge in support of their hybrid-electric vehicle developments. These strategies range from internal R&D to collaboration with external parties such as suppliers, consultants, and academic researchers.

Due to a significantly increased use of electronics and electrical components, vehicle manufacturers continually recruited electrical engineers during this time period. According to a manager at Volvo Cars, the number of engineers with electric and electronics specialization at Volvo Cars increased from about 50 in 1980 to about 1,000 in the mid-2000s [21, p. 373]. However, it is important to note that the vast majority of these engineers are active in the development of electronics and electrical components for auxiliary systems (fans, pumps, windshield wipers, rear view mirrors, and chairs with motor assisted adjustment etc.), rather than in the development of electric motors and generators for vehicle propulsion.

The input from ABB was critical for Volvo's hybrid-electric vehicle development activities during the 1990s. ABB engineers took active part in Volvo's development and demonstration projects, providing necessary expertize in electric propulsion systems. In terms of energy storage and battery technologies, Volvo lacked such a collaboration partner. Instead, Volvo established a team within their central advanced engineering division, Volvo Technology, which was devoted to energy storage and battery technologies. They, furthermore, engaged engineers from this team to evaluate battery performance and manage contacts with international battery suppliers in their hybrid-electric vehicle development and demonstration projects.

The lack of key hybrid subsystem suppliers can be considered as a weakness of the Swedish hybrid-electric vehicle TIS. During an interview, a professor active in the Swedish hybrid center admitted that their long-term R&D would have benefitted greatly from the participation of such specialized suppliers. However, it is important to note that although geographical proximity may be beneficial during explorative pre-commercial R&D activities, vehicle manufacturers select suppliers on commercial grounds rather than proximity. In order to be considered, suppliers must have product development capabilities, approved quality, reliable deliveries, global presence, and sufficient financial strength to survive and support the products [21]. Hence, having a national supplier base is not a prerequisite for the development, production, and sales of hybrid-electric vehicles. During the 2000s, both Volvo and Scania therefore collaborated with international partners for the supply of electric machinery and energy storage subsystems to their heavy hybrid vehicles.

Another example of collaboration with international suppliers is Volvo Cars' implementation of a micro-hybrid system supplied by Bosch in three of their Drive models during the late 2000s. The micro-hybrid system does not alter the vehicle power train architecture the same way as a full hybrid configuration does. Whereas a full hybrid entails a completely new architecture, the micro hybrid is more of a modular extension of a traditional ICE-based automotive power train. Therefore, the need for close collaboration to achieve integration was lower in this case. However, Volvo Cars still needed to acquire extra engineering resources, engaging international engineering consultants to understand how they could integrate the micro-hybrid system with the power train in order to benefit the most from the new technology.

The Swedish hybrid-electric vehicle TIS has repeatedly benefitted from collaboration between industry and academic researchers. One notable example is the development of the microturbine-propelled high-speed generator at KTH, a concept that ABB developed further and Volvo implemented in their environmental concept vehicles during the 1990s. Another example is how Saab Automobile engaged researchers from Lund University to support their hybrid concept and prototype developments during the early 2000s. A third example is how Volvo Powertrain employed a professor in electric machine technology as an expert consultant to support their parallel hybrid development project during the late 2000s. A fourth example is how Scania sent employees to follow PhD studies and take doctoral degrees in order to strengthen their internal engineering capabilities in areas such as electric machines. These are merely a few examples. Over the studied time period, it seems as if there has been a frequent collaboration between industry and academia.

Assessing the influence of policy, we see that subsidies from Swedish Governmental programs have filled an important role to support R&D in the Swedish hybrid vehicle TIS. In particular, these external funds made it possible for automotive managers to justify hybrid R&D activities when it was difficult to obtain internal funding. Thus, it was possible to retain engineering resources within the Swedish hybrid-electric vehicle TIS during periods characterized low legitimacy for the technology.

While the governmental programs during the 1990s were primarily focused on R&D and demonstrations, provision of engineering resources with relevant competence has become a more pronounced mission of the programs during the mid and late 2000s. The Green Car program funded PhD level courses in hybrid vehicle technology that engineers from the vehicle manufacturers could follow and the Swedish Hybrid Vehicle Center is very much focused on engineering competence development, employing several PhD students in that conduct research in technology areas such as control systems, electrical engineering, and electrochemistry. However, there is a lag in education and it seems as if it was difficult to fulfill the drastically increased need for engineering resources in the Swedish hybrid vehicle TIS during the late 1990s.

There have been several examples of industry-university collaborations in the area of hybrid vehicle technology (and related subsystem/component technologies) during the last 20 years. However, it would appear as if these collaborations primarily were relatively uncoordinated until the mid-2000s. At this stage, the establishment of the hybrid vehicle center meant that it became more institutionalized. With its emphasis on horizontal projects involving different vehicle manufacturers, it also opened up an arena for more effective resource utilization by sharing engineering resources directed at long-term and pre-competitive R&D between different manufacturers.

The lack of key subsystem suppliers in the Swedish hybrid vehicle TIS during the 2000s meant that there were no national partners available for joint hybrid technology R&D. Particularly in the area of battery technology, international manufacturers (primarily from Japan, South Korea and US) appeared to have established an impregnable lead. This deficiency of the innovation system may have delimited the possibilities for mutual long-term knowledge development, both in terms of product- and production related knowledge. In remedy, Sweden retained university-based vehicular battery technology research supported by public R&D funding. The prime motive for this was that it is important to maintain advanced battery technology knowledge in the Swedish hybrid-electric vehicle TIS to be able to serve the vehicle manufacturers with competent engineers. For the vehicle manufacturers, access to such engineering resources has been essential to be able to specify battery systems requirements and act as a professional buyer. Still, for the supply of batteries for their heavy hybrid vehicles, Volvo preferred to give Magna-Steyr an intermediate role, functioning as a battery systems integrator between Volvo and the battery suppliers. Consequently, although the support for battery technology R&D at Swedish universities may facilitate development of product-related battery knowledge, the extent to which it can facilitate sufficient development of production related knowledge in the Swedish hybrid-electric vehicle TIS seems questionable.

## 11.4 Discussion and Conclusions

### *11.4.1 The Importance of Policy*

The assessment of the Swedish hybrid-electric vehicle TIS shows how the functionality varied significantly over the period 1990–2010. The 1990s was characterized by experimentation, testing, and demonstration activities, primarily justified by a need to remedy urban air-quality problems. Governance arrangements in terms of pending regulation made it urgent for vehicle manufacturers to address these problems. The Californian ZEV-mandate probably is the most well-known example of governance arrangements in this direction. Worth noting is that among alternative propulsion technologies, battery-electric vehicles appeared as the favored alternative by regulators at this stage (early 1990s) and several large car manufacturers responded by developing, producing, and selling battery-electric cars in limited numbers. The Swedish vehicle manufacturer Volvo's focus on hybrid technology as a response to the Californian ZEV-mandate was different in this respect.

However, by the early 2000s, it had become clear that neither battery-electric nor hybrid-electric vehicle technology would not be the preferred option to solve the urban air-quality problems. The strict Californian ZEV-regulation had been softened and incremental improvements of the internal combustion engine and emission treatment technologies such as catalytic converters had significantly reduced vehicular emissions of local air pollutants. Emission regulation both in the US, Europe, and Japan continued to enforce technological progress, but this had a limited effect on hybrid technology development. Instead further improvements seemed plausible through incremental improvements and by adding more advanced emission controls to internal combustion engines (e.g. particulate filters, and systems for selective catalytic reduction and exhaust gas recirculation in the case of diesel engines).

Although there was an emerging debate on global warming in the late 1990s, with the Kyoto conference in late 1997 as a significant milestone, this did not have any major effect on the Swedish hybrid-electric vehicle TIS, at least not at this stage. Emerging activities among the Japanese competitors were not sufficient to justify any significant responses from the Swedish manufacturers. Moreover, the restructuring of ownership during this period, with GM acquiring full control of Saab Automobile, and Ford acquiring Volvo Cars reduced the Swedish manufacturers' activities in hybrid-electric vehicle technology. The level of activity in the Swedish hybrid-electric vehicle TIS thus, seems to have decreased during the early 2000s.

By the mid and late 2000s the level of activity increased significantly. The heavy vehicle manufacturers Volvo and Scania initiated efforts directed at commercialization of hybrid-electric trucks and buses. Volvo Cars developed a fuel-efficient product-line that included cars equipped with micro-hybrid systems and they announced field trials of plug-in hybrids. Saab Automobile developed hybrid prototypes, conducted field trails and displayed hybrid concept cars. Several converging landscape and industry dynamics factors promoted this increased level of activity. At this stage, the global warming debate gained momentum, the oil price had started to soar and Toyota had shown that there could be a potential business in producing hybrid-electric vehicles. The EU commission signaling a more forceful policy directed at reduced $CO_2$ emissions from vehicles served to further strengthen the motives for developing and launching hybrid-electric vehicles during this period and this was amplified by national subsidies for cars with low $CO_2$ emissions in several European countries, as well as in Japan. Still the financial crisis and collapsing markets in the late 2000s put the vehicle manufacturers in a very difficult situation.

On a Swedish level, two notable governance arrangements are the KFB program for development and demonstration of electric and hybrid vehicles during the 1990s and the Green Car programs that offered R&D subsidies for (amongst others) hybrid-electric vehicle technology during the 2000s. These two programs were both commissioned by the Swedish Government, and both can be characterized as R&D subsidy programs. Still these programs are very different in terms of the instruments and mechanisms they deployed and the functions of the TIS that they addressed. With its involvement of local public authorities and focus on demonstrations, the KFB program of the 1990s was very public. It addressed a broad set of issues; supporting both technology oriented R&D and social science research (market surveys and studies on the impact of the technology on the society). Hence, knowledge development and diffusion in a broad sense was a primary concern. However, the KFB program also addressed market formation. In fact, one of the program goals was to support the introduction and demonstration of electric and hybrid vehicles. In particular, the program (as well as a parallel public procurement program) was designed to stimulate market demand through empowerment of potential customers. By supporting potential customers, an early market for electric and hybrid vehicles was supposed to emerge, a market that would be attractive for vehicle manufacturers to enter. However, in the end this "demand pull" arrangement was deemed ineffective. In retrospect, actors in the program appear to have been overly optimistic about how mature the technology was. Moreover, they overestimated their possibilities to influence and in the end the program failed to engage the car manufacturers. By contrast, the heavy vehicle manufacturers received more direct support and according to the evaluation report, these parts of the program were also more successful [15].

The Green Car programs during the 2000s addressed a more exclusive set of actors than the preceding KFB program. These public–private partnerships primarily involved the manufacturers themselves, as well as researchers at technical universities within relevant fields of technology. The Green Car programs were designed to fit explicit needs as expressed by the Swedish vehicle manufacturers, and thus these programs also became less public then the preceding KFB program.

Moreover, they became more narrowly oriented, first and foremost focusing the development and diffusion of product-related knowledge, and to some extent resource mobilization through education of engineers. The design of the Green Car programs can be seen as a reaction to observed shortcomings of the KFB program in terms of weak involvement from the vehicle industry. Letting the vehicle manufacturers orchestrate the program was a means to ensure a strong industrial participation. An evaluation of the Green Car program also maintains that the program was successful in facilitating R&D projects and interaction between different vehicle manufacturers, and it was also successful in establishing networks between industry and academia in relevant fields of technology [22].

However, a well-functioning technological innovation system has to support several subprocesses of innovation. From a TIS-perspective, it is therefore possible to question the narrow focus on the development of product-related technological knowledge. Technology development is merely one part of it; technological innovation has to be considered in a broader context. Interview respondents also acclaim that development of hybrid-electric vehicles in the Swedish TIS often has been slowed down by deficient organizations rather than by technological shortcomings. This refers both to the vehicle manufacturers' internal processes for assessing and understanding market developments and needs, as a basis for strategic decisions on product development, as well as processes for acquiring new knowledge and establishing efficient production networks involving new suppliers. In the early 2000s, leading Japanese hybrid car manufacturers had already started climbing the learning curve rapidly in terms of production and market related knowledge. The steeply increased sales during the mid-2000s further reinforced this and additional manufacturers entered the market. In terms of hybrid car production, Toyota appeared to have a clear lead by the late 2000s, which Honda as an obvious runner-up. Other car manufacturers (including the Swedish ones) were left behind. In the heavy vehicle sector, the Swedish manufacturers were better positioned, much due to the fact that no global manufacturer so far had attained the position as an international leader in hybrid vehicle technology. Volvo announced ambitions to become such a leader, but at by the end of 2009 their release of a hybrid truck on the market had been postponed. Still, they maintained their ambitions, introducing a hybrid bus to the market in 2010. However, still after 20 years of R&D in the Swedish hybrid-electric vehicle TIS, no Swedish hybrid car was available on the market and no significant volumes of heavy hybrid vehicles had been produced. The reluctance to enter the market severely restricted the possibilities to generate production and market-related knowledge. However, in terms of Swedish governance arrangements, little was done to explicitly address these issues during the 2000s.

Having this stated, it is important to note is that it appears very difficult to influence global vehicle manufacturers by implementing governance arrangements on a national level. Experiences from the KFB program (as well as the parallel Nutek public procurement program) illustrate that this may be particularly true for relatively small nations like Sweden with a very limited domestic market. The case study shows how the Swedish hybrid-electric vehicle TIS was very much influenced by industrial dynamics, international competition and global trends such as the oil price, the financial situation, and international debates on emissions from road transport (both in terms of NOx, PM and $CO_2$). The case further shows that strong signals were required to legitimate hybrid R&D and particularly strong signals from several sources were required for the manufacturers to initiate product development aiming for market introduction. To be able to take these decisions, the vehicle manufacturers had to justify them internally in their respective organizations. Hence, they needed to be confident that a market for these vehicles would develop and that this market would be large enough to justify the required investment. With a limited domestic market, the search for such signals will primarily be pursued on the international scene.

## 11.4.2 Implications for Technology Management in Firms

From a management perspective, the case study of hybrid-electric vehicle technology in Sweden 1990–2010 provides a neat illustration of the complexity associated with innovation for environmental sustainability in the auto industry. The case points at the extensive time needed for development and validation of new technologies. Only by the end of the period, the technology was deemed ripe for market introduction. These decisions on technology commercialization had been preceded by extensive testing, experimentation and validation activities, involving both various technical configurations and a multitude of product applications. In the end, it would appear as if the firm with the broadest product portfolio was best positioned to introduce hybrid-electric vehicle technology to the market. This is because a broad portfolio makes it more likely that the firm will be able to find a suitable niche application, which facilitates market entry. However, the transition from such a niche strategy to a more widespread adoption of the technology may still constitute a critical step for the firm.

The case analysis shows how the demand side of innovation for environmental sustainability is characterized by multiple and disparate drivers. Demands for environmental performance and support for environmental innovation often come from different sources than other demands, and the relationships between different actors and their initiatives are often difficult to predict. Moreover, it is important to consider global trends in terms of e.g. fuel prices, environmental debates, and finance. Decisions as to whom to interact with therefore seem critical.

Interaction between manufacturers and users has often been emphasized in the innovation-management literature as an important factor supporting effective innovation. However, such interactive learning processes may take different forms and involve different actors in cases of innovation for sustainability. The case of hybrid-electric vehicle technology in Sweden 1990–2010 specifically emphasizes the need for manufacturing firms, which are promoting innovation for environmental sustainability, to interact with both users and policy makers at a number of different levels. In an analysis of various policy instruments that stimulate innovation for sustainability, Chapman [23] suggests that regulators and policy makers need to learn about innovative processes within industrial firms. An inverse proposition may be directed toward industrial managers; namely, that effective sustainable innovation assumes that affected managers learn more about policy processes.

## References

1. Nelson RR, Winter SG (1982) An evolutionary theory of economic change. Harvard University Press, Cambridge
2. Pierre J, Peters G (2006) Governing complex societies. Palgrave Macmillan, Basingstoke
3. Hekkert MP, Suurs S, Smits REHM (2007) Functions of innovation systems: a new approach for analysing technological change. Technol Forecast Soc Chang 74(4):413–432
4. Bergek A, Jacobsson S, Carlsson B, Lindmark S, Rickne A (2008) Analyzing the functional dynamics of technological innovation systems: a scheme of analysis. Res Policy 37(3):37–59
5. Magnusson T (2011) Hybrid-electric vehicle technology in Sweden 1990–2010. Report to Stockholm Environment Institute, Stockholm
6. BIL Sweden (2009) Bilismen i Sverige. Årsbok BIL Sweden
7. Norgren L, Nilsson R, Perez E, Pohl H, Sandström A, Sandgren P (2007) Needs-driven R&D programmes in sectoral innovaiton systems. Vinnova Anal VA 2007:15
8. Faguert S, Arnold E, Eriksson M-L, Jansson T, Mattsson P, Niklasson L, Salino P, Segerpalm H, Åström T (2009) Effekter av statligt stöd till fordonsforskning: betydelsen av forskning och förnyelse för den svenska fordonsindustrins konkurrenskraft. Vinnova Anal VA 2009:02
9. Takeishi A (2002) Knowledge partitioning in the interfirm division of labor: the case of automotive product development. Organ Sci 13(3):321–338
10. Clark KB, Fujimoto T (1991) Product development performance: strategy, organization, and management in the world auto industry. Harvard Business School Press, Boston
11. Fogelberg H (2000) Electrifying visions: the technopolitics of electric cars in California and Sweden during the 1990s. Doctoral thesis Göteborg University Science and Technology Studies
12. Bauner D (2007) Towards a sustainable automotive industry—experiences from the development of emission control systems. Doctoral thesis KTH
13. Pohl H (2010) Interparadigmatic hybrids—A new theoretical concept when addressing paradigmatic shifts in technology. Int J Innov Technol Manag 7(4):353–375
14. Overgaard L, Folkesson A (2007) Scania hybrid concept—with robust technology into the future. In: 57th UITP World congress, pp 1–14
15. Hedman J, Laage-Hellman J, Lindqvist E (2000) Utvärdering av KFBs el- och hybridfordonsprogram. KFB-Inf 2000:27
16. Kemp R, Schot J, Hoogma R (1998) Regime shifts to sustainability through processes of niche formation: the approach of strategic niche management. Technol Anal Strateg Manag 10:175–195
17. Von Hippel E (1988) The sources of innovation. Oxford University Press, New York
18. Andersson BA, Jacobsson S (2000) Monitoring and assessing technology choice: the case of solar cells. Energy Policy 28:1037–1049
19. Sperling D, Gordon D (2009) Two billion cars. Oxford University Press, Oxford
20. Turrentine TS, Kurani KS (2007) Car buyers and fuel economy? Energy Policy 35:1213–1223
21. Eriksson S (2007) Electrical machine development—a study of different machine types from a Swedish perspective. Doctoral thesis KTH
22. Faguert S, Arnold E, Eriksson M-L, Jansson T, Segerpalm H, Thoresson-Hallgren I, Åström T (2007) Samverkan for uthållig konkurrenskraft—Utvärdering av fordonsforskningsprogrammet Gröna Bilen. Faugert & Co Utvärdering
23. Anex RP (2000) Stimulating innovation in green technology: policy alternatives and opportunities. Am Behav Scientist 44(2):188–212

# Chapter 12
# Utilizing Risk Analysis and Scenario Planning for Technology Roadmapping

**Yuya Kajikawa, Yasunori Kikuchi, Yasuhiro Fukushima and Michihisa Koyama**

**Abstract** Technology roadmap aims to direct future paths of advancing technologies and to foster understandings among stakeholders. The path illustrated in the roadmap is regarded as agreed commitment on the targeted technology and enables selected investment on it. However, current technology roadmaps are limited in its adequacy and reliability, because roadmapping tends to focus on proinnovation perspective rather than precaution one. Risks and uncertainties embedded in the presented pathways and major consequences in the future realized

Y. Kajikawa (✉)
Graduate School of Innovation Management, Tokyo Institute of Technology, 3-3-6 Shibaura, Minato-ku, Tokyo 108-0023, Japan
e-mail: kajikawa@mot.titech.ac.jp

Y. Kikuchi
Presidential Endowed Chair for "Platinum Society", the University of Tokyo, 7-3-1 Hongo, Bunkyo-ku, Tokyo 113-0033, Japan

Y. Fukushima
Department of Environmental Engineering, National Cheng Kung University, No. 1 University Road, Tainan, Taiwan 701, China

Y. Fukushima
Sustainable Environment Research Center, National Cheng Kung University, Taiwan, No. 500, Sec. 3, An-Ming Road, Annan District, Tainan, Taiwan 70955, China

Y. Fukushima
Research Center for Energy Technology and Strategy, National Cheng Kung University, No. 1 University Road, Tainan, Taiwan 701, China

M. Koyama
Inamori Frontier Research Center, Kyushu University, 744, Motooka Nishi-ku, Fukuoka 819-0395, Japan

M. Koyama
International Institute for Carbon-Neutral Energy Research, Kyusyu University, 744, Motooka Nishi-ku, Fukuoka 819-0395, Japan

D. Cetindamar et al. (eds.), *Strategic Planning Decisions in the High Tech Industry*, DOI: 10.1007/978-1-4471-4887-6_12,

by following the roadmaps are often not well presented. In this chapter, we propose to utilize risk analysis and scenario planning for resolving risks and uncertainties in technology roadmap. The approach includes identification of technological, economic, environmental and social risks, and constraints. It also utilizes scenario planning to reduce uncertainties of the future. We discuss the feasibility of the approach in a case of roadmapping of energy technologies.

## 12.1 Introduction

### *12.1.1 What is Technology Roadmap?*

Science and technology (S&T) roadmap is emerging as an indispensable management tool in the modern society, because it can direct future paths of advancing technologies and foster understandings among stakeholders including scientists, engineers, research and development (R&D) managers, and policy makers. Currently, roadmap is widely adopted in industry, government, and academia for providing a way to develop technology strategy, identify gaps and opportunities in R&D, and plan for resource allocation. Such a rising importance of S&T roadmap reflects complexity, specialization, and the rapid pace of development in global R&D activities, which make S&T management difficult. These also request managers to invest to specific and selected target and to orchestrate related efforts and stakeholders.

S&T roadmap is usually regarded as "a consensus articulation of a scientifically informed vision of attractive technology futures" [1]. "A 'roadmap' is an extended look at the future of a chosen field of inquiry composed from the collective knowledge and imagination of the brightest drivers of change in that field" [2, 3]. Roadmaps are both forecasts of what is possible or likely to happen, as well as plans that articulate a course of action [4]. In short, roadmaps are defined as the views of a group of stakeholders as to how to get where they want to go to achieve their desired objective [5]. It is expected to navigate stakeholders to the way illustrated in the roadmap and to act as a clue for orchestration of R&D activities and strategies.

Roadmapping is normative approach and includes back casting from vision to present status along the pathways to realize the vision [6]. In contrast, forecasting is descriptive approach based on retrospective data from past to present. By extrapolating current trends to the future, we predict future trends and obtain a prospective perspective. These two approaches are not contradictory but compatible. Vision is usually given by a top-down approach based on the expert's experience and intuition, but the feasibility should be tested against the existing data and trends. Therefore, scientifically informed vision included in the Branscomb's definition on roadmap should be based on the forecasted future of scientific output. In this sense, technological forecasting is one of key components for S&T roadmapping to construct a reliable roadmap.

The definition and characteristics of S&T roadmap are analogous to that of a road map [7]. In everyday life, a road map is a layout of the paths or routes of some particular geographical space. It is abstracted representation of reality illustrated for specific purpose. Roadmap is aimed to be used by travelers to select among alternative routes in determining how to arrive at a particular destination. S&T roadmaps are usually represented by a time-directed two-dimensional layout including markets, products, and technologies [8–10]. Therefore, S&T roadmap can give us a guideline to decide where to walk in market, product, and technology space.

One of the differences between roadmap and S&T roadmap is the existence of commitment. Although travelers can choose a route they want to walk, S&T roadmap is rather a represented result of the route that stakeholders have selected among plausible routes about S&T. Consensus and commitment on S&T roadmap are regarded even as the critical factors of successful roadmaps [7]. In sum, S&T roadmap is not only a consensus articulation of attractive technology futures but also a means and an articulation of means to realize attractive technology futures. Actually, in some S&T roadmaps, we can see that the elements like talents, skills, projects, and budgets are explicitly mentioned to link these resources with S&T, which are needed to realize the path illustrated in the roadmap. Therefore, S&T roadmap is subjective and strategic but neither objective nor value neutral.

Utilization of S&T roadmap and selected investment might lose opportunity. But it is becoming inevitable. Previous studies have empirically demonstrated the positive relationship between the investments and the outcomes of scientific research [11, 12]. But in the circumstance where total budget is constrained or has declined [13], we cannot bet for all options, while it is not a rudimentary task for R&D managers and policy makers to make decisions on effective investment in promising and emerging technologies by selecting them from a pile of plausible candidates. Concentrated and collaborative investment under budget constraints on a specific research target enhances the possibility of that technology to be realized in society. In this context, Robert Galvin describes a roadmap as becoming an inventory of possibilities in a particular field, thus stimulating more targeted investigations [2, 3].

## *12.1.2 Typology of Roadmap*

S&T roadmap has been successfully implemented and utilized in semiconductor industry. And nowadays it is used in a variety of sectors like materials, energy, software, transport, healthcare, information and communication technology (ICT), and so on. However, application capability and feasibility in other sectors are not well-investigated yet. Yasunaga et al. [14] proposed three categories of S&T roadmap having different driving force; market-pull, technology push, and social-needs-driven.

The first category is market-pull roadmap, where the primary flow of the roadmapping idea descends from the market, to the product and then to technology like a roadmap in semiconductor industry. In this case, there is a motivation and expectations for stakeholders to work for realizing roadmap because of the pressure of visible market trend. In this category, roadmap is co-created among rather well-defined stakeholders like suppliers and customers.

The second one is technology-push roadmap like nanotechnology roadmap. In this case, when technology shown in the roadmap is realized, it is promising that they open new market or substitute the existing market. We do not always have a clear market image but convince the existence of market when technology is ready or disseminated in market even under the development phase. Therefore, stakeholders also have a motivation and motivation to realize technologies in market and society. In this type, researchers in universities and research universities and engineers in firms take an initiative in roadmapping.

The third is social-needs-driven roadmap like environmental and renewable energy technology, where future market and its trend is uncertain. Additionally, large externality exists and coverage of technology is broad. Therefore, there is no clear driving force and motivation for each stakeholder in the current state to spend efforts in high-quality roadmapping and commitment on the roadmap. In this category, it is usually difficult to comprehend an entire system and reach a consensus on roadmap. Roadmap tends to be a patchwork of roadmaps about technology components. And more seriously, risk associated in the focal technology is unintentionally missed or intentionally neglected. Due to the subdivision of speciality and task, it is difficult to check reliability and feasibility of roadmap. And there is an 'interest gap'. One benefiting from technology differs from one suffering from it. Therefore, it needs institutions to compensate the gap. When there is an 'interest gap', it becomes harder to notice and assess the risk embedded in the path of the roadmap.

## 12.2 Theoretical Background

### *12.2.1 Risk Analysis*

In the existing roadmaps, plausible risks are not usually stated in a comprehensive and explicit manner, while attractive technology futures are well visualized. It means that these roadmaps recommend bright future paths without claiming risks embedded in those paths. Stirling [15] addresses technology appraisal and choice by contrasting proinnovation and precaution attitude toward technology in policy process.

Researchers have widely advocated that technical, commercial, organizational, and social risks and uncertainties must be considered in technology roadmapping [10] and future technology assessment [16, 17]. The process of risk analysis, risk assessment, and risk management consists of the following steps [17]: (1) scope definition; (2) hazard/risk identification; (3) hazard/risk estimation, probability, and consequences; (4) risk evaluation; (5) proposals for actions; (6) risk reduction/control. Risk analysis includes the steps from (1) to (3). Risk assessment is (1–4), and risk management is (1–6). The similar type of procedures is also reported in [18].

But in practical roadmapping processes, less effort has been paid to identify, analyze, and state risks than to claim the brightest and attractive technology futures. Risk analysis and assessment is needed to functionalize social-needs-driven roadmap. Stirling [19] and Leach et al. [20] distinguished risk, uncertainty, ambiguity, and ignorance. Factors categorized into risk are not problematic in knowledge about both outcomes and likelihoods. We know what happens with quantitative probability. Uncertainty means a status where probability of an event is unknown, while we know such an event will occur. In ambiguity, probability of the occurrence of an event is known, but there is no agreement on interpretation, selection, measurement, and prioritization of outcome of the event. In ignorance, both outcome and likelihood are unknown.

For example, temporal variation of electricity supply by installation of wind turbine and solar cell is considered as risk, because we can know the probability distribution function of energy supply by them with considering the past meteorological data at the place. There is a book reporting Wind Turbine Syndrome, which is self-reported symptoms, incl., sleep disturbance, headache, tinnitus (ringing in the ears), and the others. However, underlying mechanism and a portion of people suffering from such syndrome are obscure [21]. Therefore, health effect of wind turbine is uncertain. People may consider that influence of offshore wind power on fishery has ambiguity. It might disturb ecosystem in the coast and affected livelihood of fishers and spill-over effect to the residential in that region. In some cost estimation of nuclear power, we can see that capital and social costs are ignored. According to Severance [22], capital cost occupies the largest portion in projected total electricity generation cost.

We might not be able to classify elements relating with a certain technology into risk, uncertainty, ambiguity, and ignorance *a priori*. It depends on framing of issues, existing and accessible knowledge, interpretations, and validation boundary. But these classifications can be used to know the current status, recognition and perception of stakeholders and promote risk analysis, risk assessment, and risk management.

However, the above argument has the critical drawbacks. That is the argument focuses on knowledge about outcome caused by introducing technology, but not input and causality [23]. And more importantly, they are time dependent, and future is in nature uncertain and ambiguous, which requests us to introduce scenario planning approach. When technological and social status of a technology is given, we can explore, identify, and assess risks, but they will change and are not determined in advance before future comes. Therefore, we need assumptions about future, i.e., scenario.

### 12.2.2 Scenario Planning

The uncertainty of the future can be evaluated and resolved across a number of scenarios which share plausible futures. Durance and Godet [24] defined scenario as the following, "A scenario is not a future reality but rather a means to represent it with the aim of clarifying present action in light of possible and desirable futures". Scenario is also defined as "an internally consistent view of what the future might turn out to be—not a forecast, but one possible future outcome" [25]. Well-written scenarios have the following characteristics [26]: internally consistent; link historical and present events with hypothetical events in the future; carry storylines that can be expressed in simple diagrams; plausible; reflect predetermined elements; identify signposts or indicators that a given story is occurring.

Three types of scenarios, normative, predictive, and exploratory scenarios, are distinguished and they are used for different purposes [24, 27, 28].

Normative scenario is constructed from alternative images of the future which may be both desirable and feared, and are conceived in a retro-projective way. They respond to the question "how can a specific target be reached?" Therefore, normative scenarios are similar to roadmap, and scenarios that take their starting point in one or several well-defined targets.

Predictive scenario analyzes past and present trends and presumes likely future. It is similar to forecasting but also adopts what-if approach. The aim to utilize predictive scenario is to answer 'what will happen?' Forecasting will illustrate future scenario with extrapolation from past to present and answer 'What will happen provided that the most likely development unfolds?' Predictive scenario responds to the question, 'what will happen, on the condition of some specified event?'

Exploratory scenario differs from forecasting because of its exploratory purpose. It will illustrate "what can happen". Compared to predictive scenario, exploratory scenario planning is rather external and strategic. Exploratory scenario analyzes external factors to the focal issue, while predictive one does future trend of the issue itself. The aim of strategic scenarios is to describe a range of possible consequences of strategic decisions. They respond to the question "what can happen if we act in a certain way?" Therefore, we can consider that risk analysis is a part of exercise of exploratory scenario planning.

Predictive scenario analyzes factors affecting feasibility of roadmap, and exploratory one assesses factors affected if future illustrated in the roadmap is realized. Predictive and exploratory scenario planning exercise is expected to deal with discontinuities and surprise factors, which opening-up alternative view about future [29]. As suggested by Strauss and Radnor [30], roadmapping can become unwieldy when planning is challenged by change that is volatile and rapid, systemic and unanticipated. To overcome this limitation, strategic thinking and scenario planning highlighting the implications of possible future and systemic discontinuities should be used to prepare unexpected event and its consequence. Exploratory scenario planning has a potential to provide operational and strategic-level support for managers by opening-up broader technological, social, and environmental contexts and perspectives. It also assists to identify, assess, and manage uncertainties and risks. And it contributes to test the robustness of proposed technology roadmaps and influence the wider policy context [29, 31].

## 12.3 Roadmapping of Energy Technologies

### *12.3.1 Case Description*

We conducted a participatory case study to confirm effectiveness of risk analysis and scenario planning in roadmapping. The authors engaged in roadmapping process by division of energy engineering (DEE) at the society of chemical engineers Japan (SCEJ) as editorial board members of the roadmap from 2009–2010. DEE published their first energy roadmap in 2005 [32]. The roadmap which the authors contributed to is the second edition published in 2010. The roadmap covers a wide range of energy technologies including coal, nuclear power, fuel cell, biomass, heat pump, and so on. The board members invited the contributors illustrating each technology, and contributors prepared their chapters. Total 34 contributors are from universities, 19 are from industries, and 7 are from public institutes. Board members requested each contributor to illustrate a roadmap on the technology where they are specialist. Board members and contributors had two meetings in 2009 and also two meetings in 2010. During the meetings, board members explained agenda and aim of the roadmapping. Board members also had intensive workshops two times in 2009 and five times in 2010. In the following, we briefly explain our approach for risk analysis and scenario planning.

In the roadmapping, we asked each contributor to identify risk relating with that technology. However, it is not a rudimentary task for each contributor to identify risk without any framework. In this study, we first define the scope of risk and then explain our scope to each contributor. The scope includes the following factors: (1) materials (toxicity, resource scarcity, uneven distribution, etc.); (2) technological feasibility (possibility to be realized); (3) competitive technology; (4) necessary related technology (e.g., refinement, waste treatment, etc.); (5) infrastructure; (6) manufacturing capacity; (7) byproduct; (8) others (social acceptance, etc.). We asked contributors to consider these factors.

We need future scenario and integrated modeling for explicitly showing social benefits that we will have when the scenario described in the roadmap is realized. And integrated modeling is necessary to count the broad scope of technologies. In this chapter, we estimated $CO_2$ reduction potential in integrated scenarios of energy technologies. Plausible scenario (Scenario A) and expected reduction scenario (Scenario B) were developed in the workshops by board members. Scenario B is more proactive to adapt renewable energy technologies and energy-saving technologies like biomass, high temperature gas reactor, electrification of residential, and electric vehicle.

**Table 12.1** Future scenarios in electricity generation sector

| | Share of power source | | |
|---|---|---|---|
| | Base case (%) | Scenario A (%) | Scenario B (%) |
| Thermal power (coal) | 25.2 | 5.2 | 2.8 |
| Thermal power (oil) | 13.3 | 2.0 | 2.0 |
| Thermal power (LNG) | 13.6 | 0.0 | 0.0 |
| Thermal power (LNG-combined) | 13.6 | 20.9 | 20.9 |
| SOFC-GT | 0.0 | 7.0 | 7.0 |
| Nuclear power | 25.4 | 43.0 | 43.0 |
| Hydraulic power | 7.9 | 8.3 | 8.3 |
| Photovoltaic | 0.4 | 5.5 | 5.5 |
| Wind power | 0.4 | 4.4 | 4.4 |
| Geothermal power | 0.3 | 0.9 | 0.9 |
| Biomass | 0.0 | 0.3 | 2.8 |
| Waste power and biomass | 0.0 | 2.4 | 2.4 |

We also analyzed the structure of energy use and production in 2007 as a benchmark (Base case). For the analysis of data at 2007, we used the data of Agency for Natural Resources and Energy [33] and Greenhouse gas Inventory office of Japan [34]. The main parameters in these scenarios assumed in this study are summarized in Tables 12.1, 12.2, and 12.3. Energy efficiency of each energy technology is adopted by each roadmap described by contributors. Other parameters include dissemination ratio of SOFC into residential and offices, ratio of

**Table 12.2** Future scenarios in residential, offices and industrial sector

| Technology scenario | | Scenario A | Scenario B |
|---|---|---|---|
| Dissemination of SOFC into residential | Ratio | 0 % | 60 % |
| Electrification of kitchen in residential | Ratio | 30 % | 100 % |
| Electrification of heating in residential | Ratio | 100 % | 100 % |
| Dissemination of SOFC into office | Ratio | 0 % | 100 % |
| Electrification of kitchen in office | Ratio | 30 % | 30 % |
| Electrification of heating in office | Ratio | 50 % | 50 % |
| Energy input in agricultural sector | Ratio(2007)[a] | 22.5 % | 22.5 % |
| Dissemination of SOFC into pulp, chemical, ceramic, and machinery industries | Ratio | 25 % | 25 % |
| Energy input change by utilizing low-temperature exhaust heat | Ratio(2007) | 76.9 % | 76.9 % |
| Energy input change by chemical heat pump | Ratio(2007) | 96.2 % | 96.2 % |
| Energy input change by membrane for separation | Ratio(2007) | 82.0 % | 82.0 % |
| $CO_2$ emission in steel industry | Ratio(2007) | 70.0 % | 70.0 % |
| SOFC-GT in thermal power (LNG) | Ratio | 25 % | 25 % |
| High-temperature gas reactor | Plant number | 2 | 3 |
| Power generation from biomass | Ratio (Coal) | 5.0 % | 5.0 % |

[a] Ratio (2007) is relative performance compared to that at 2007

Table 12.3 Future scenarios in transport sector

| | Fuel consumption | Share | | | |
|---|---|---|---|---|---|
| | MJ/km | Long distance | | Short distance | |
| | | Scenario A (%) | Scenario B (%) | Scenario A (%) | Scenario B (%) |
| Gasoline vehicle | 2.1 | | | | |
| Diesel vehicle | 1.695 | 95.0 | 0.0 | 10.0 | 0.0 |
| Natural gas vehicle | 2.1 | | | | |
| Gasoline hybrid vehicle | 1.475 | | | | |
| Diesel hybrid vehicle | 0.945 | | | | |
| Natural gas hybrid vehicle | 1.695 | 0.0 | 0.0 | 75.0 | 0.0 |
| Fuel cell hybrid vehicle | 0.550 | 5.0 | 0.0 | 7.5 | 0.0 |
| Electric vehicle | 0.868 | 0.0 | 100.0 | 7.5 | 100.0 |

electrification in residential, and so on. With these scenarios and parameters, we estimated the social return by estimating the reduction of $CO_2$ emission.

In Table 12.1, we assumed large share of energy supply by nuclear power plant, which is based on energy basic strategy (EBS) approved by the Japanese Cabinet at June 2010. In EBS, share of nuclear power in electricity supply was aimed to reach at 53 % in 2030, while that at 2007 is 26 %. We must note that after FUKUSHIMA incident caused by earthquake and tsunami at 11 March 2011, the EBS was abandoned, and energy mix, especially the share of nuclear power, is still active and prior social and policy issue at August 2012 when the authors wrote this chapter. Although this chapter assumed the share as shown in Table 12.1, it must be reanalyzed to reflect and respond social needs and opinions after FUKUSHIMA and will be reported in another chapter. We already published a chapter which analyzes and proposes very short-term countermeasures after FUKUSHIMA to avoid power outage [35]. But the chapter does not analyze long-term transformation of energy system in Japan. In this chapter, we do not analyze risk and scenario about nuclear power.

## 12.4 Results and Implications

In the following, we will show a brief result of the roadmapping on energy technologies, especially focusing on risk analysis and assessment of $CO_2$ reduction potential.

By introducing risk analysis, we identified a variety of risks embedded in roadmaps. Most of risks identified by the contributors were material risk due to the scarcity of materials. Examples are Li, Fe, Co, and Ni for lithium ion battery, Pt for proton exchange membrane fuel cell (PEMFC), Pt and Ni for chemical heat pump, and so on. This frequent notion to material risk might be due to the fact that contributors of each roadmap are mainly engineers. We must note that whether material risk appears or not depend on scenario about dissemination ratio assumed in Table 12.2. When each technology is widely disseminated, risk of resource scarcity will increase. Therefore, it is confirmed that scenario planning is important for risk analysis and assessment.

Another important category of risk was competitive technology. Examples are capacitor for lithium ion battery, secondary battery and heat pump for PEMFC, PEMFC for direct methanol FC, and so on. These complicated relationships of competitive technologies make energy roadmapping difficult. One of the function of S&T roadmap is to share future vision on a selected technology among alternatives, which enables more concentrated effort on the selected target. However, in energy technologies, it is not easy to determine a specific technology as the target to be pursued. This is because of uncertainty of future outcome by R&D. Another reason is that one single technology cannot solve the multiple issues including, economic efficiency, energy security, environment, and safety. For example, even when solar cell occupies an essential part of energy system, it cannot cover all energy source due to the constraints like fluctuation of energy supply and required large land area. We must utilize different technologies like secondary batteries, chemical storage, and gas combined cycle. And dissemination of solar cell is influenced by policy initiatives like feed-in-tariff and cost of the other technologies. Future scenario of an energy technology, therefore, depends on the others.

Requirement of infrastructure is also well recognized. Examples are hydrogen station for FC, smart grid for renewable, forest road for biomass, exhaust heat collection for chemical heat pump, and so on. These infrastructures require large investment by public or private sector. Roadmap and decision making on infrastructure are essential factor for dissemination of energy technology and transition to a new energy system. This requests roadmapping of an energy technology including planning of investment to infrastructure.

As shown in the above, contributors of roadmapping well-identified risks in materials and competitive technology. However, about the rest of factors in our risk category, we had fewer responses than expected. We had less answers on technological feasibility, necessary related technology, and manufacturing capacity. About necessary related technology, the reason is broadness of scope of energy technologies. Diverse sectors are related with the development of energy technologies from energy, electronics, chemistry, materials, ICT, and so on. Many contributors just stated that the risk in technological feasibility is their cost, although they lack detailed and reliable cost analysis. Roadmapping based on reliable cost analysis and strategy to decrease risks, but it is remaining issues for our energy roadmaps.

Figure 12.1 shows the result of our analysis about current and future $CO_2$-emission based on scenario described before. We can see current emission of 1,300 Mt/year will be decreased by 35.2 % in scenario A and 44.4 % in scenario B. In scenario B, we assume larger introduction of biomass as an energy resource, and therefore $CO_2$-emission in energy transformation increases due to less transformation efficiency of biomass. In scenario B, we also assume all vehicles are electronic vehicles, and therefore we need more energy to feed electricity for them. But increase in energy transformation is compensated for less energy use in transport.

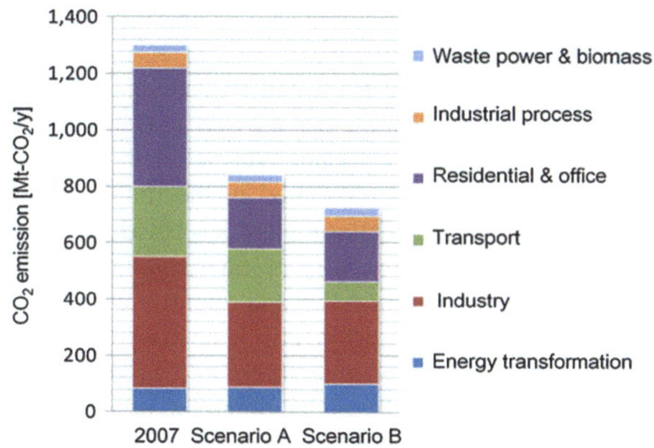


**Fig. 12.1** $CO_2$ reduction based on future scenarios

We performed sensitivity analysis on $CO_2$-emission by changing parameters used in these scenarios, and found that share of power source has a substantial impact on the result of the assessment for each technology. The composition of power source even affects the contribution of each technology to the reduction of $CO_2$-emission. It means that it is difficult for each technology like FC and heat pump to assess their own impact only by analyzing those focal technologies. Even when we select a specific technology target and illustrate a roadmap of a technology, we need predictive scenario about the other technology and also energy system to assess the outcome of that technology. As already discussed, there is a typology of scenario: predictive, normative, and explorative scenario. Our scenarios, i.e., scenario A and scenario B, are predictive one, and roadmap of each technology is a normative scenario. In this study, we independently set scenario A and scenario B from roadmaps of individual energy technologies. However, they should have been concurrent, because of dependency and complex relationships among energy technologies and entire system. One approach is to utilize normative scenarios (roadmap) as predictive scenarios for the other roadmaps. Another approach is to transform normative scenarios by utilizing the other roadmaps and predictive scenarios as exploratory ones. One can test outcome and sensitivity of each roadmap by integrated modeling as shown in this chapter, which has a possibility to work as exploratory scenario planning tool. This interaction between scenario planning, risk analysis, and roadmapping is needed especially for energy roadmap, and will be future direction of research in S&T roadmapping.

## 12.5 Concluding Remarks: Suggestions to Managers

Technology roadmap is an indispensable management tool especially for R&D managers and policy makers. In the existing roadmaps, although attractive technology futures are well illustrated, plausible risks and uncertainties relating with development and deployment of technology are not usually stated in a comprehensive and explicit manner.

In this chapter, we proposed risk analysis and scenario planning as tools for technology roadmapping. We showed a brief result of our roadmapping activity on energy technologies in Japan. We included risk analysis and scenario planning to identify uncertainties associated in the roadmaps and plausible outcome of $CO_2$ reduction when the technologies described in the roadmaps are realized. Inputs from more than 40 experts in energy sector were integrated and assessed to explicitly state inherent risks, and quantify expected outcomes. However, we must note that risk identification is not an easy task because of cognitive and social factors. We also found that integrated modeling of technologies are crucial for reliable assessment. Findings obtained in the execution of this case study are organized as useful guidelines to construct a useful, reliable, and informative technology roadmap.

From our experience, concurrent approach of scenario planning, risk analysis, and roadmapping, is crucial, because of dependency and complex relationships among energy technologies and entire system. Another important issue, which is not discussed well in this chapter, is role of contributors and facilitators in roadmapping and importance of community. Community created during roadmapping is the key resource. We must give more attention on epistemic community sharing norm, knowledge, validity, and task [36]. Collective minds, trust, dynamic tension embedded in academic society and between wide varieties of societies are vehicles driving broader-based process of technology appraisal. As stated by Stirling [15], it is also necessarily to involve "active and transparent stakeholder deliberation and citizen participation. This is not just a necessity for the sake of legitimacy in the final stages of technology choice. It is also crucial to the initial 'framing' of procedures for scientific and technical assessment". This interactive and concurrent approach will be future direction of research in S&T roadmapping.

## References

1. Branscomb LW, Keller JH (1998) Towards a research and innovation policy. In: Branscomb LW (ed) Investing in innovation; creating a research and innovation policy that works. MIT Press, Cambridge
2. Galvin R (1998) Science roadmaps. Science 280:803
3. Galvin R (2004) Roadmapping-a practitioner's update. Technol Forecast Soc Change 71: 101–103
4. Kappel TA (2001) Perspectives on roadmaps; how organizations talk about the future. J Product Innov Manag 18:39–50
5. Probert D, Radnor M (2003) Technology roadmapping. Res Technol Manag 46:27–30
6. McDowall W, Eames M (2006) Forecasts, scenarios, visions, backcasts and roadmaps to the hydrogen economy: a review of the hydrogen futures literature. Energy Pol 34:1236–1250
7. Kostoff RN, Schaller RR (2001) Science and technology roadmaps. IEEE Trans Eng Manag 48:132–143

8. Phaal R, Farrukh CJP, Probert DR (2004) Technology roadmapping-a planning framework for evolution and revolution. Technol Forecast Soc Change 71:5–26
9. Rinne M (2004) Technology roadmaps; infrastructure for innovation. Technol Forecast Soc Change 71:67–80
10. Petrick IJ, Echols AE (2004) Technology roadmapping in review; a tool for making sustainable new product development decisions. Technol Forecast Soc Change 71:81–100
11. Rosenberg N (1974) Science, invention, and economic growth. Econ J 84:90–108
12. Narin F, Hamilton KS (1996) Bibliometric performance measures. Scientometrics 36:293–310
13. Nemet GF, Kammen DM (2007) U.S. energy research and development; declining investment, increasing need, and the feasibility of expansion. Ene Pol 35:746–755
14. Yasunaga Y, Watanabe M, Korenaga M (2009) Application of technology roadmaps to governmental innovation policy for promoting technology convergence. Technol Forecast Soc Change 76:61–79
15. Stirling A (2007) Deliberate futures: precaution and progress in social choice of sustainable technology. Sustain Dev 15:286–295
16. Hall JK, Martin JC (2005) Value-added chain; a framework for evaluating radical technology development. R&D Manag 35:273–284
17. Koivisto R, Wessberg N, Eerola A, Ahlqvist T, Kivisaari S, Myllyoja J, Halonen M (2009) Integrating future-oriented technology analysis and risk assessment methodologies. Technol Forecast Soc Change 76:1163–1176
18. Power M, McCarty LS (1998) A comparative analysis of environmental risk assessment/ risk management framework. Env Sci Technolo 32:224–231
19. Stirling A (2008) Science, precaution, and the politics of technological risk: converging implications in evolutionary and social scientific perspectives. Ann NY Acad Sci 1128:95–110
20. Leach M, Scoones I, Stirling A (2010) Dynamic sustainabilities: technology, environment, social justice. Earthcan, London
21. Knopper LD, Ollson AC (2011) Health effects and wind turbines: a review of the literature. Environ Health 10:78
22. Severance CA (2009) Business risks to utilities as new nuclear power costs escalate. Electricity J 22:112–120
23. Aven T (2011) On different types of uncertainties in the context of the precautionary principle. Risk Anal 31:1515–1525
24. Durance P, Godet M (2010) Scenario building: uses and abuses. Technol Forecast Soc Change 77:1488–1492
25. Porter ME (1998) Competitive advantage. Free Press, New York
26. Van der Heijden K (1996) Scenarios: the art of strategic conversation. Wiley, New York
27. Börjeson L, Höjer M, Dreborg K-H, Ekvall T, Finnveden G (2006) Scenario types and techniques e towards a user's guide. Futures 34:723–739
28. Höjer M, Ahlroth S, Dreborg KH, Ekvall T, Finnveden G, Hjelm O, Hochschorner E, Nilsson M, Palm V (2008) Scenarios in selected tools for environmental systems analysis. J Cleaner Prod 16:1958–1970
29. Volkery A, Ribeiro T (2009) Scenario planning in public policy: understanding use, impacts and the role of institutional context factors. Technol Forecast Soc Change 76:1198–1207
30. Strauss JD, Radnor M (2004) Roadmapping for dynamic and uncertain environments. Res Technol Manag 47:51–57
31. Saritas O, Aylen J (2010) Using scenarios for roadmapping: the case of clean production. Technol Forecast Soc Change 77:1061–1075
32. Kato Y (2007) Innovative energy materials from Honebuto energy road map. J Chem Eng Jpn 40:1141–1149
33. Agency for natural resources and energy (2010) http://www.enecho.meti.go.jp/info/statistics/index.htm. (Accessed April, 2010)
34. Greenhouse gas inventory office of Japan (2010) http://www-gio.nies.go.jp/index-j.html. (Accessed April 2010)

35. Fukushima Y, Kikuchi Y, Kajikawa Y, Kubota M, Nakagaki T, Matsukata M, Kato Y, Koyama M (2011) Tackling power outages in Japan: the earthquake compels a swift transformation of the power supply. J Chem Eng Jpn 44:365–369
36. Haas PM (1992) Introduction: epistemic communities and international policy coordination. Intl Organ 46:1–35

# Chapter 13
# Application of Six Sigma Methodology and Use of Data in Decision Making in the Fixed Line Planning Process

**Gordan Krčelić, Siniša Štifanić and Milan Meden**

**Abstract** On today's market of continuous technology advancements, increased globalization and constant competitive pressure, companies mostly rely on information systems and automation of business processes and their improvements. Timely and correct decision making is essential in every business environment, and the telecommunication industry in particular it is in urgent need of capital investment and development plan resolutions. Moreover, the complexity of the processes in the telecommunication industry requires a concept of their improvement and an approach to achieving operational excellence which is recognized in the Six Sigma methodology. The objective of this work is to show operational improvement of a fixed network planning process derived from Six Sigma knowledge, methods, and tools. The Model for quantification of benefits for the given investment in access network and possible ways to identify the best locations for investment based on conclusions drawn from raw data will be presented.

## 13.1 Introduction

The development of new technologies along with telecommunication market trends makes demands on telecommunication service providers (Telcos) to offer high speed Internet access and advanced interactive services within their portfolio of

G. Krčelić (✉) · S. Štifanić · M. Meden
HT—Croatian Telecom/Telecom Infrastructure Services, 10000 Zagreb, Croatia
e-mail: gordan.krcelic@t.ht.hr

S. Štifanić
e-mail: sinisa.stifanic@t.ht.hr

M. Meden
e-mail: milan.meden@t.ht.hr

D. Cetindamar et al. (eds.), *Strategic Planning Decisions in the High Tech Industry*, DOI: 10.1007/978-1-4471-4887-6_13,

services. Video, as part of the advanced triple play service already exists on almost all European markets while cable network providers are securing a market for Video on Demand (VoD) and High-Definition TeleVision (HDTV) as well. Telcos realized a long time ago that the battle is worth fighting, and in order to remain competitive in that niche, they should provide sufficient bandwidth. The trends show precisely which direction Telcos should pursue which is additionally supported by the National Strategy of Broadband access issued by the Government of the Republic of Croatia [1]. The National strategy covers a time frame from 2011 till 2015 and has its starting point in the European Union (EU) Agenda for a digital Europe.

The availability of high speed and ultra-high speed broadband access in national strategies of the countries Members of the European Union refers mostly to 100 % coverage of the population with access speeds from 20 to 100 Mbit/s until the end of 2015. The National strategy encourages a balanced development of broadband Internet access; the implementation of an effective principle to reduce the digital gap between Croatian regions as well as the digital gap between the Republic of Croatia and the EU Member States.

To offer the bandwidth required by advanced services and to satisfy customers' expectations, special attention should be paid to access network which, at present, is not ready to support future requirements. Given the need for high speed Internet access, a possible option is to invest into optical access network. Investment in the development of fiber optical (FO) access network substantial to cover the whole territory of Croatia requires enormous financial and technical resources as well as huge manpower effort but also implies open regulatory issues. The other option is to focus on shortening the copper subscriber loop by installing the Very-high-bit rate Digital Subscriber Line (VDSL) node which would provide sufficient speed adequate to provide advanced service quality. Such approach would save time until the FO access reaches every customer.

The article describes possible concepts of access network development, the application of Six Sigma methodology on the access network planning process and the Model for estimations of profitability in access network reconstruction projects. At the end, it will be shown how the Model itself can be used to compare various concepts of access network reconstruction, and also how to use data to make business decisions.

## 13.2 Relation Between Subscriber Loop Distance and Available VDSL2 Speed

### *13.2.1 VDSL2 Standard*

The VDSL2 standard was adopted by the International Telecommunication Union (ITU) in May 2006. It was designed to enable copper networks to achieve speeds greater than 25 Mbit/s at mid-subscriber loops (up to 1,800 m) and short loops (up to 300 m) speed close to 100 Mbit/s for symmetrical data transmission.

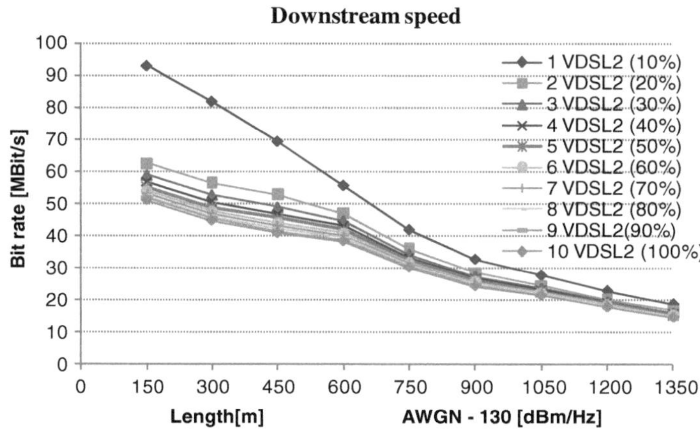


**Fig. 13.1** VDSL2 Bit rate for the 8b profile depending on the percentage of penetration

The VDSL2 standard contains most of the functionalities of ADSL2+ standard (Asymmetric Digital Subscriber Line), including advanced diagnostics, a joint management interface, and the ability to maximize available bandwidth and data rates which make it an ideal access technology for providing video content.

To simplify the task of configuring network equipment, the VDSL2 standard defines profiles tailored for different regional deployment architectures such as the Central Office (CO), remote Digital Subscriber Line Access Multiplexer (DSLAMs) or the Fibre-To-The-Cabinet (FTTC), the digital loop carriers and multi-dwelling units. There are eight profiles which define the power options from 11.5 to 20.5 dBm, the bandwidth up to 30 MHz and a minimum data rate for each profile [2].

Figure 13.1 shows the data rates for the 8b profile depending on the percentage of penetration, which is ideally suited for longer reach applications such as CO or FTTC deployments with long loop access network.

### *13.2.2 Accessed Network Deployment Scenarios*

To accommodate the bandwidth requirements of triple play and interactive services, service providers are extending the fibre deeper into the access network. Besides the Fibre-To-The-Home (FTTH), there is a hybrid approach by which the fibre is fed with a remote DSLAM or remote node in the FTTC concept, shortening the copper connection to the home under 1,500 m. As shown in Fig 13.2, this approach exploits further the existing copper infrastructure and bridges the bandwidth gap between fiber and copper deployment concepts while decreasing the cost and time needed to deploy the fiber all the way to the premises [2].

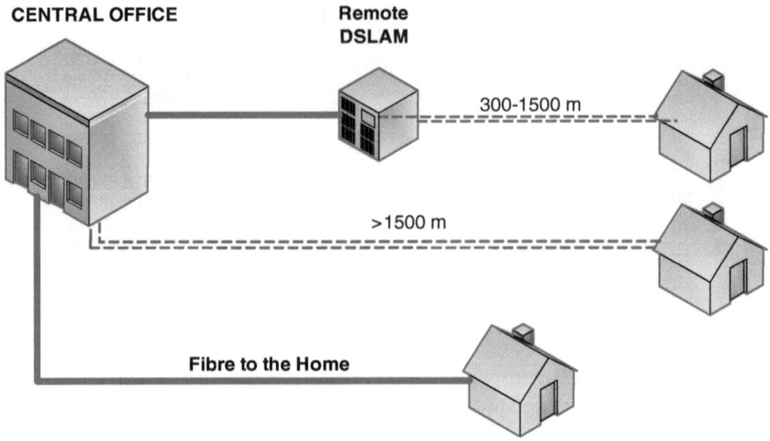


**Fig. 13.2** Typical access network scenarios

## 13.3 Reaching Operational Excellence Through Six Sigma

### *13.3.1 Introducing Six Sigma*

To enhance competitiveness, companies have embraced a variety of different management programs over the last two decades. One of them is the Six Sigma ($6\sigma$) methodology. It focuses on the improvement of complex business processes in order to reduce variations and solve the process problems along with the business problems [3]. To apply the $6\sigma$ management philosophy means to take a strategic and disciplined, customer, and process-centred and data-driven approach to performance improvement. Its main objective is to improve the processes to meet or exceed customers' requirements and enable organizations to operate better, faster, at a lower cost, and to achieve operational excellence. The $6\sigma$ achieves service performance through effective utilization of nonstatistical and powerful statistical tools and techniques of the Define, Measure, Analyze, Improve, and Control (DMAIC) cycle.

### *13.3.2 Applying Six Sigma in a Company*

Once all the business participants are aware of the implementation of the measuring and controlling system as a strategic baseline of the company, this can be considered as a well-performed preparation for the implementation of the $6\sigma$ methodology. The complexity of the processes in the telecommunication industry reveals many potential areas for improvement.

The $6\sigma$ projects need to be carefully selected, planned, and reviewed to maximize the benefits of their deliverables. In order to do so, the projects themselves have to be feasible, organizationally and financially beneficial, and customer oriented [4].

As the available technology is rapidly moving forward, telecommunication trends impose the need for new investments in the existing access network. Those potential investments put the network planning process under direct spotlight since the new services require a new investment cycle and now, more than ever, it is important to focus on a structured approach. Companies recognized the $6\sigma$ method as the right approach to improving the planning process and achieving the most from the invested.

### *13.3.3 Applying Six Sigma to the Access Network Planning Process: VoC and the Definiton of CTQ and CTB in a Company*

The key factor of the $6\sigma$ project success is in its ability to link to the customer. Accordingly, the starting point for the DMAIC methodology is to go through the "Voice of the Customer (VoC)" in the Define phase. Here, the most important thing is to recognize the customers and identify their "true" needs [5]. A clear set of measures and metrics has to be defined to incorporate customers' requirements [6]. The Critical to Quality (CTQ) uses the VoC and converts it into a clear and measurable language.

According to the Enhanced Telecom Operations Map (e-TOM), the output of the network planning process (S.3.1 Resource Strategy and Planning) [7] is oriented toward internal and external customers. External customers are service users while the internal ones are recognized in the sales and marketing department. Internal customers are in the focus of this project. Their requirements on the planning process deliverables can be expressed as, *I want each customer to have sufficient attainable speed to sell both the existing and future services*, which is in fact the CTQ according to the $6\sigma$ methodology.

As the $6\sigma$ is determined to meet the customers' needs profitably on the one hand, on the other hand, it is necessary to identify the "Voice of the Business (VoB)", and therefore define the Critical to Business (CTB) as a way to improve efficiency in business processes. The business objective states that the available budget should be used in the most profitable way and that is considered as CTB.

The business objective has to be measurable which raises the question as to how projects can be compared on the basis of profitability. Each project is unique in its scope, technology, benefits, budget, and so on. The challenge is to seek for the data-driven method that would help prioritize the most profitable ones. Two approaches are considered:

- The decision-making framework applicable for a large-scale, multi-participant, multi-criteria decision analysis such as the Analytical Hierarchy Process (AHP) [8], or
- The development of a Model not only able to rank the projects but also to show the profitability of each.

Since the output of the AHP is a dimensionless number in the range between zero and one, and it does not give any information on the ROI, it has been decided to develop a Model that would satisfy the requirements for the fulfillment of a business objective.

## 13.4 The Model for Calculating Investment Profitability in Copper Access Network

### *13.4.1 Vision of the Model*

On the one side, there are the CTQ and CTB, and on the other side the needed budget related to the scope of each project. After the Define phase, according to the DMAIC approach, the next phase is Measure. Its aim is to collect data to see that the CTQ is met, and ensure that the measurements are accurate. Then the data are used to produce graphs and charts to present the process problems visually [5].

Since the geographical scope of the projects can vary according to the reconstruction of the whole access network, one cable, several distribution points, or just one distribution point (DP), it is necessary to determine the potential benefits of each project. If we have insight into the maximal potential benefits of the given investment, and if we are able to measure how much the investment in the access network infrastructure influences the identified potential, and if we can estimate the required budget, then we can calculate their relationship. If we do that for each project, we can unambiguously compare them and sort them according to that fraction.

Conceptually, the Model has been designed as shown in Fig. 13.3. The potential benefit is calculated according to the data from the corporate Data Warehouse System (DWH) which collects data from the Customer Related Management system (CRM), Billing, Operational Support Systems/Business Support Systems (OSS/BSS), and Resources Databases. Two distinct sources for the calculation of the potential benefits are identified as:

- The upselling potential of the existing services, *ESp*
- The potential arising from prevention of income erosion as a comparison of the current throughput of each network segment and the predicted distribution of income by ranges of speed in 5 years, *IEPp*.

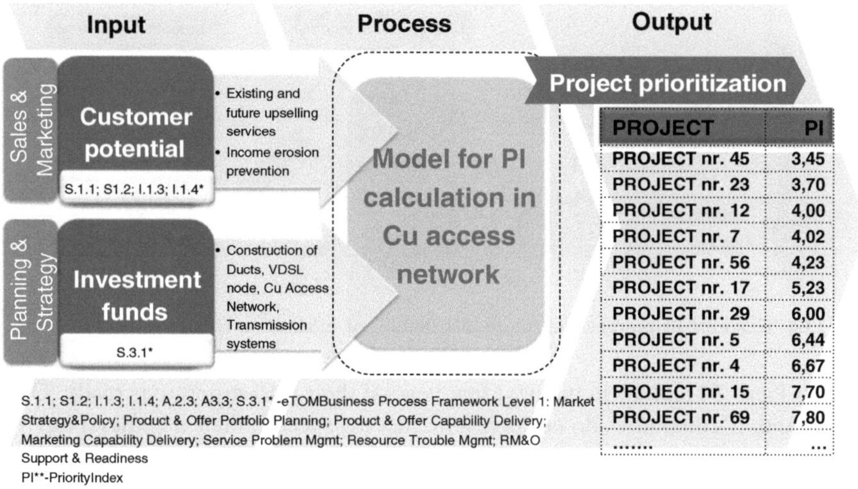


**Fig. 13.3** Model for calculating profitability of investment in copper access network

The upselling potential of the existing services represents the possibility to earn income from new customers in the given area selling them the existing services if there are sufficient technical capabilities. If the resources were not the limitation factor, the Waiting List (WL) would not exist. The WL is therefore a collection of customers' requests that could not be met.

In the area where Telcos have WLs, there is usually much greater potential to sell the existing services than it is visible from the WL itself. Assuming that the population living along one cable has similar telecommunication habits and needs, then it is further presumed that the existing broadband services have equally penetrated into all the DPs on that cable. If we calculate the penetration into the DPs with technical possibilities and apply those data to the DPs without them, we can identify new potential customers and calculate the additional revenue for the existing upselling services.

The potential of preventing income erosion represents the current income in a given area to be protected. The Telcos objective is to increase their income or at least to maintain it at the same level. Since the customers' demands for video and Internet speed are rising, new services need to be launched and they will require more bandwidth, while at the same time the prices for the existing services will drop. If the investment into the access network fails to happen, the income is doomed to go down and the customers' base is likely to downsize. In order to measure the risk of income erosion, all the DPs in the network are classified into six ranges, called Tiers, defined by the available speed they can provide the end-users with. Nowadays, Telcos generate most of their incomes from speeds lower than 20 Mbit/s as shown in Fig. 13.4.

The governments of some countries put pressure on resource providers to develop the infrastructure that would provide future services with speeds higher than 20 Mbit/s, or more [1]. It is, therefore, expected that the Telcos income will

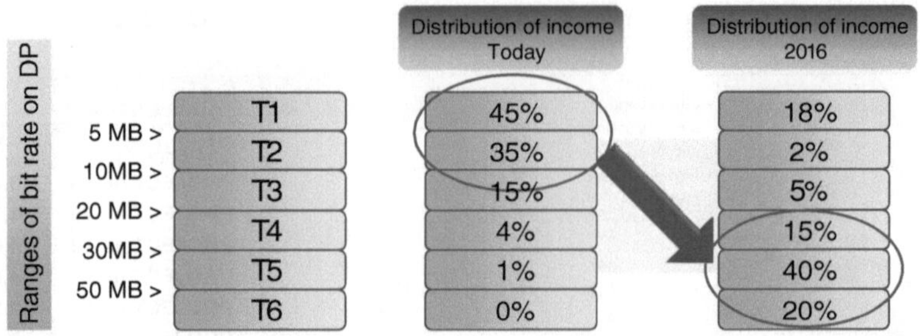


**Fig. 13.4** Example of present and future distribution of income on DPs in respect to Tiers

shift to Tiers T4 and T5 in the next few years if their infrastructure is ready. On the other hand, if companies do not invest in "access speed", their income, if any, will be more and more at risk.

### *13.4.2 Use of Data to Calculate the Benefits of Reconstruction Projects in Access Network Vision of the Model*

As already mentioned, the geographical scope of reconstruction projects in the access network can vary on a large scale, so that the minimal project entity has to be defined—large enough not to lose the variation in the area and small enough not to be too detailed. We have chosen a DP as the smallest network unit, and all the potential benefits are calculated on the DP level as the sum of potentials of all the customers connected to that DP. The potential benefit of a project is in this way the sum of potential benefits of all the DPs in a particular project scope, e.g.

$$Bp = \sum_{i=1}^{m} \left( DPp_{ESp_i} + DPp_{IEPp_i} \right) \qquad (13.1)$$

where:

| | |
|---|---|
| $Bp$ | Annual benefit potential for the given project scope [€]; |
| $m$ | {1, 2, 3,…} total number of DPs in the project scope; |
| $DPp_{ESPi}$ | Annual upselling potential of existing services on *ith DP in* [€]; |
| $DPp_{IEPi}$ | Annual prevention potential of income erosion on *ith DP in* [€]. |

To meet the CTB requirements, we have to compare the projects with various scopes, the budget and potential benefits, and determine which projects are better than the others. It is not necessary to calculate the ROI itself for each of those projects (although it could also be done), but just to gain an insight into it. Consequently, we have introduced a dimensionless number called the Priority Index

($PI$). The benefits of the $6\sigma$ projects are presented as annual values, so the $PI$ roughly represents a ROI for each project and is calculated as follows:

$$PI_n = \frac{\sum_{i=1}^{m} (DPp_{ESp_i} k_{DPES_i} + DPp_{IEPp_i} k_{DPIEP_i})}{I_n} \quad (13.2)$$

where:

| | |
|---|---|
| $PI_n$ | Priority Index for the project scope "$n$" [-]; |
| $n$ | {1, 2, 3...} number of the project; |
| $m$ | Total number of DPs in the project scope "$n$"; |
| $k_{DPESi}$ | $\{0 \leq k_{DPESi} \leq 1\}$ Influence factor for the existing upselling services on $DP_{PESPi}$; |
| $k_{DPIEPi}$ | $\{0 \leq k_{DPIEPi} \leq 1\}$ Influence factor for prevention of income erosion on $DP_{PIEPi}$; |
| $I_n$ | Investment budget [€] Required for the project scope "$n$". |

It is essential to stress the importance of the influence factor on the maximal annual benefit potential for each DP in the given scope. Investing in an area with high potential does not necessarily provide a high return on investment and the influence factor defines that relation. The smaller the $PI$ is, the shorter the ROI, the better the project. At the end, there will be a prioritized list of development projects in the access network in respect to the $PI$. The total available budget will then be distributed to the projects on the list identifying the top profitable ones.

### 13.4.3 Estimation of Future Available Speed Using VDSL2 Standard

On the basis of the introduced Model, we have identified the maximal benefits for each DP providing the available access rate per DP is greater than 50 Mbit/s as in Tier 6. However, with the installation of the VDSL node to an existing CO location or a new location as FTTC, the deployment will not affect equally the available bit rate on all the DPs of a particular network or cable. The impact depends on a lot of factors and the most important ones are: the local loop distance, the twisted pair chord diameter, and the type of copper cable, the quality of joints and penetration of BB services.

To determine the function of the local loop distance, penetration and available speeds for ADSL2+ and VDSL standards, we used the linear regression to measurement data collected in the testing environment on the most common type of cable in the access network as shown in Fig. 13.1 [5]. To calculate the attainable speed applying the ADSL2+ standard, we used the following expression:

$$DPs_{A_i} = C_A + \beta_{A1} d_{A_i} + \beta_{A2} p_i \quad (13.3)$$

where:

| | |
|---|---|
| $DPs_{A_i}$ | available ADSL2+ speed on *i*th *DP* [Mbit/s]; |
| $C_A$, $\beta_{A1}$, $\beta_{A2}$ | Regression parameters; |
| $d_{Ai}$ | Electrical distance between the ADSL node and *i*th *DP*; |
| $p_i$ | Penetration of broadband servicesin the basic group. |

For the VDSL2 standard we considered only the 50 % penetration curve (see Fig. 13.1) which we set as a minimal requirement the future constructed network should fulfill:

$$DPs_{V_i} = C_V + \beta_V d_{V_i} \tag{13.4}$$

where:

| | |
|---|---|
| $DPs_{V_i}$ | Future available VDSL2 speed on *i*th *DP* [Mbit/s]; |
| $C_V$, $\beta_V$ | Regression parameters; |
| $d_{Vi}$ | Electrical distance between the VDSL2 node and *i*th *DP*. |

In both cases the determination coefficients $r^2$ were above 95 % $\left(r^2_{\text{ADSL}} = 98\,\% \text{ and } r^2_{\text{VDSL}} = 96\,\%\right)$ which confirm that the linear regression model approximates the empirical values well.

To determine the distance between the future VDSL2 node and the DP, their geo-coordinates stored in the resource database were used. For a more accurate estimate of the future electrical distances, the resulting air distances are multiplied by a factor calculated as a fraction of the estimated electrical distances between each DP and the existing node on the cable (stored as an attribute in the database). This factor is called the "electrical distance factor", $k_{de}$.

As mentioned earlier, the distance is not the only variable influencing the available speed on the DP. We have introduced an additional corrective factor called "other influences factor", $k_o$ which represents all the other variables influencing the available speed on the DP such as the distance of customers from the DP, the quality of joints and insulation of the cable segment, specific electrical interferences, and so on. The other influences factor is calculated as a fraction of the median of the existing ADSL2+ speeds on each DP and the calculated attainable speed according to the expression:

$$k_{o_i} = \frac{\tilde{C}s_{A_i}}{DPs_{A_i}} \tag{13.5}$$

So for the future available speed at each DP it is:

$$DPs_{V_i} = C_V + \beta_V k_{de} k_{o_i} \sqrt{(x_N - x_{DP})^2 + (y_N - y_{DP})^2} \tag{13.6}$$

where:

| | |
|---|---|
| $\tilde{C}s_{A_i}$ | Median of ADSL2+ measured speeds for all customers on ith *DP* [Mbit/s]; |

| | |
|---|---|
| $DPs_{A_i}$ | Calculated attainable ADSL2+ speed on ith *DP* [Mbit/s] for the existing distance from the node; |
| $DPs_{V_i}$ | Future available VDSL speed on ith *DP* [Mbit/s]; |
| $k_{oi}$ | other influences factor [-]; |
| $x_N$, $y_N$, $x_{DP}$, $y_{DP}$ | Geographical coordinates of the node and *DP*; |
| $k_{de}$ | Electrical distance factor [-]; |
| $C_V, \beta_V$ | Regression parameters. |

### 13.4.4 Determination of the Budget Required to Install the VDSL Nodes

To establish the profitability of each investment and therefore sort all of the projects according to their profitability, it is important, in addition to their benefit estimation, to estimate the investment amount as well. Conceptually, the new VDSL node can be installed either in:

- The existing ADSL2+ node location, or
- A new location.

If the VDSL node installation takes place at the existing ADSL2+ CO, provided that there is enough technical space to install a new rack, enough power supply and air conditioning capacity, with an already installed uplink, then the cost can be calculated as follows:

$$In_{\mathrm{VCO}} = n_R R + n_p P \tag{13.7}$$

where:

| | |
|---|---|
| $In_{\mathrm{VCO}}$ | Investment cost for VDSL node in the central location CO [€]; |
| $R$ | Cost of the rack [€] |
| $P$ | Cost of the Printed Circuit Board with VDSL2 *ports* [€]; |
| $n_R$ | Number of racks; |
| $n_P$ | Number of PCB's. |

The number of racks and Printed Circuit Boards (PCB) is determined by the number of customers that will be connected to the node.

If the VDSL node installation takes place at a new location, except for the cost of the cabinet and the PCBs, there are the costs of connecting the node to the back bone network uplink which depends on the availability of ducts and the cost of the transmission system:

$$In_{\mathrm{VFTTC}} = n_C C_C + d(C_{CP} + pC_d) \tag{13.8}$$

where:

| | |
|---|---|
| $In_{\text{FTTC}}$ | Investment cost for the VDSL node at a new FTTC location [€]; |
| $n_C$ | Number of FTTC cabinets; |
| $C_C$ | Cost of one FTTC cabinet [€]; |
| $d$ | Distance from the new node to the back-bone network [$m$]; |
| $C_{CP}$ | Cost of cable pulling [€/$m$]; |
| $p$ | Percentage of the distance "d" where it is necessary to construct cable ducts [%]; |
| $C_d$ | Digging cost [€/$m$]. |

For the expressions (13.7, 13.8) we have used the data collected from the ADSL node installation projects carried out in the access network in the past years.

### 13.4.5 Implementation of the Model in the Access Network Planning process

Following the Model for estimating the benefits in a certain scope and the expressions for determining the required budget, the decision-making process at operational level has improved significantly. Furthermore, from the $6\sigma$ point of view, the Model itself represents a system for measuring quality in the planning process and offers possible solutions for the "vital few" root causes of not choosing the right project among the access network investment projects discovered in the analysis phase. After having verified the assumptions built in the Model, we have implemented the Model along with other solutions in the planning process.

Besides, if we use the Model for simulating the data about all the DPs in the network, the strategic investment and development processes [e-TOM: Resource Strategy and Planning (S.3.1) and Resource Capability Delivery (S3.2)] [7, 9] can be improved as well. Thus, it is possible to provide a data-driven answer to questions such as:

- What are the maximal possible benefits, and what funds are required to install the VDSL node if we choose the existing CO or FTTC deployment?
- Which are the most profitable sites for investment in the access network?
- How many VDSL nodes are required and what funds do we need if we want all the customers in the copper access network to have minimal attainable speed of 30 Mbit/s? And so on.

The development of simulations which could give an answer to those questions exceeds the scope of this article. In the following lines, we will focus on the first question and provide an answer to it on a randomly selected sample of 150 existing nodes which contain 786 cables with 15.231 DPs in order to reduce the time to process the data for the whole population. The following assumptions were considered:

- Only the existing DP locations were taken into consideration as possible locations for a new node.
- If the VDSL node is installed on a CO location, this influences the identified potential on all the DPs and all the cables connected to the node.
- If we install the VDSL node on the existing CO location, this will affect half of the identified maximal upselling potential of the existing services, ($ESp = 50\ \%$) if the DPs jump to the next Tier.
- If we install the VDSL node on a new location (FTTC), this will affect all the identified maximal upselling potential of the existing services, ($ESp = 100\ \%$) if the DPs jump to the next Tier.
- In the FTTC case, only one VDSL node with the smallest *PI* would be installed on each cable.
- To calculate the budget required for the installation of a new VDSL node, $p = 20\ \%$ of the network is lacking in cable ducts.

The simulation has provided us, among many others, with the following conclusions:

- If we install the VDSL nodes only on CO locations, this can affect only 28 % of the identified *ESp* potential and 40 % of the *IEPp* potential.
- If we install the VDSL node as a FTTC deployment, the impact ratio rises to 86 % of the *ESp* potential and 78 % of the *IEPp* potential.
- If we install the VDSL nodes on a CO, 33 % of the locations will have a *PI* less than 5, with those locations covering 35 % of the maximal identified potential.
- If we compare the benefits from the investment required to install the VDSL node in the 5 % best locations in a FTTC deployment with the benefits that would result from investing the same amount in average locations, the benefit ratio is 9:1 which confirms the importance of choosing the right locations to invest in.

To get a broader picture, we can put the *PI* values on $x$-axis and the influence on cumulative attainable potentials on $y$-axis as shown in Fig. 13.5.

If the main objective of the Telco is to reduce the risk of income erosion, then it is better to install the VDSL nodes on the existing CO locations in the first phase of investment (values with $PI < 6$). However, if it is more important for the Telco to generate additional revenue by means of selling the existing services, then a better solution would be to install new VDSL nodes as the FTTC deployment although the *PI* in that period is slightly worse, but after the sixth year the benefits are significantly higher.

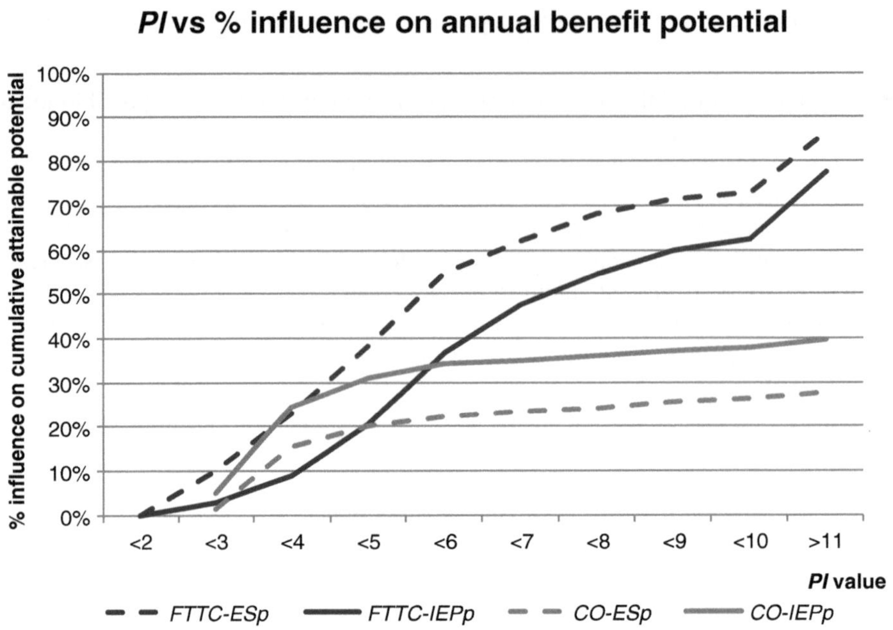


**Fig. 13.5** VDSL node installation on CO versus FTTC deployment in respect to *PI* and the influence on Potential of Income Erosion ($IEP_p$) and the upselling potential of the Existing Services ($ES_p$)

## 13.5 Conclusion

The application of the 6$\sigma$ methodology on the access network planning process helped us recognize true customers' needs and the planning process weaknesses through the VoC, VoB, and the use of the company's data. In the measure phase, it was necessary to install a measurement system and so we decided to develop a Model. In the analysis phase, after having gathered important information and identified areas to improve, we have realized that the Model itself turns out to be the solution to the "vital few" root causes of not choosing the most profitable project in the Improve phase of the DMAIC cycle.

Through the use of simulation we have compared the VDSL CO and FTTC deployment scenarios and determined the best locations to invest in. Although the Model itself and the conducted simulation are based on assumptions, the picture they create can be different from reality. This discrepancy can be reduced with further reality checks and fine tuning of assumptions.

Following the principles behind the proposed Model and estimations of the investment cost for various concepts of telecommunication network development, the impact of strategic decisions can be studied. Furthermore, the Model can be upgraded to consider the operational costs for network maintenance, which is directly connected to locations with poor network quality. So the strategy should not be oriented only to sales, growth, and the prevention of income erosion. It can also be oriented to reducing the cost of network maintenance. The size of impact on the reduction of maintenance cost depends on the concept of selected network reconstruction.

Besides having correct and quality decisions based on the results obtained from the simulation and correct assumptions for the simulation, it is also essential to:

- Ensure the quality (being correct and up to date, having integrity) of the collected data used to calculate the DP potential in the Model.
- Adjust constantly the assumptions used in the Model.
- Be able to interpret the results obtained from the simulation in the right way.

The article shows that the $6\sigma$ methodology combined with data analysis can boost process excellence and decision making at both the operational and strategic level.

## References

1. Government of the Republic of Croatia (2011) Strategy for broadband development in the Republic of Croatia for 2012–2015. Available at http://www.mppi.hr/UserDocsImages/w%20StrategyBB%205_12.pdf
2. Aware Inc. (2006) VDSL2 The ideal access technology for delivering video services. White paper revision 2
3. Nave D (2002) How to compare six sigma, lean and the theory of constraints. Qual Prog 35(3):73–78
4. Anbari F, Kwak YH (2004) Success factors in managing six sigma projects. Project management institute research conference, London, 11–14 July
5. Lunau S, Meran R, Roenpage O, Staudter C (2008) Six sigma$^{+Lean}$ toolset. Executing improvement projects successfully. Springer and UMS, Germany
6. Anthony J, Banuelas R (2002) Key ingredients for the effective implementation of six sigma program. Measuring Bus Excellence 6(4):23
7. TM Forum (2009). Business process framework (eTOM). Document GB921D v8 available at http://www.amdocs.com/public/etom8.pdf
8. Saaty TL (2008) Decision making with the analytic hierarchy process. Int J Serv Sci 1(1):83–98
9. Fiammante M (2010) Dynamic SOA and BPM best practices for business process management and SOA agility. International Business Machines Corporation, From Simplified Integration to Dynamic Processes:4–6

MIX
Papier aus verantwortungsvollen Quellen
Paper from responsible sources
FSC® C105338

If you have any concerns about our products,
you can contact us on
**ProductSafety@springernature.com**

In case Publisher is established outside the EU,
the EU authorized representative is:
**Springer Nature Customer Service Center GmbH**
**Europaplatz 3, 69115 Heidelberg, Germany**

Printed by Libri Plureos GmbH
in Hamburg, Germany